SpringerBriefs in Applied Sciences and Technology

Series editor

Janusz Kacprzyk, Polish Academy of Sciences, Systems Research Institute, Warsaw, Poland

SpringerBriefs present concise summaries of cutting-edge research and practical applications across a wide spectrum of fields. Featuring compact volumes of 50–125 pages, the series covers a range of content from professional to academic.

Typical publications can be:

- A timely report of state-of-the art methods
- An introduction to or a manual for the application of mathematical or computer techniques
- A bridge between new research results, as published in journal articles
- A snapshot of a hot or emerging topic
- An in-depth case study
- A presentation of core concepts that students must understand in order to make independent contributions

SpringerBriefs are characterized by fast, global electronic dissemination, standard publishing contracts, standardized manuscript preparation and formatting guidelines, and expedited production schedules.

On the one hand, **SpringerBriefs in Applied Sciences and Technology** are devoted to the publication of fundamentals and applications within the different classical engineering disciplines as well as in interdisciplinary fields that recently emerged between these areas. On the other hand, as the boundary separating fundamental research and applied technology is more and more dissolving, this series is particularly open to trans-disciplinary topics between fundamental science and engineering.

Indexed by EI-Compendex and Springerlink.

More information about this series at http://www.springer.com/series/8884

Mohd Firdaus Yhaya · Husnul Azan Tajarudin
Mardiana Idayu Ahmad

Renewable and Sustainable Materials in Green Technology

Mohd Firdaus Yhaya
Bioresource, Paper and Coatings Technology Division, School of Industrial Technology
Universiti Sains Malaysia
Penang
Malaysia

Mardiana Idayu Ahmad
Environmental Technology Division, School of Industrial Technology
Universiti Sains Malaysia
Penang
Malaysia

Husnul Azan Tajarudin
Bioprocess Technology Division, School of Industrial Technology
Universiti Sains Malaysia
Penang
Malaysia

ISSN 2191-530X ISSN 2191-5318 (electronic)
SpringerBriefs in Applied Sciences and Technology
ISBN 978-3-319-75120-7 ISBN 978-3-319-75121-4 (eBook)
https://doi.org/10.1007/978-3-319-75121-4

Library of Congress Control Number: 2018933480

Printed on acid-free paper

This Springer imprint is published by the registered company Springer International Publishing AG part of Springer Nature
The registered company address is: Gewerbestrasse 11, 6330 Cham, Switzerland

Preface

This book aims to present a full picture of balanced and refreshing views as well as the state-of-the-art research on renewable and sustainable materials, from green technology perspective. It emphasises on fundamental processes and productions of renewable and sustainable materials from chemical and biological approaches as well as their various applications in green technology field. It also presents up-to-date reviews, researches and methods which are outlined in an organised way to ensure that the links of multidisciplinary approaches in the area of engineering, technology, and applied sciences are achieved.

The book is arranged in didactic writing style with the aim to provide a guide and insight into existing knowledge of renewable and sustainable materials for contemporary teaching and research, aided with diagrams. It is divided into five main chapters in which every chapter incorporates in-depth technical information without isolation of theory and reality. It also includes key technological development and advancements in the area of renewable and sustainable materials in relation to green technology for solving human and environmental issues in the modern era leading to a greener future. In addition, challenges, future outlook, and opportunities of these renewable and sustainable materials are presented and discussed. In Chap. 1, the book outlines a compendiary abridgment of important keywords within the scope of this book with the aim to provide a foundation for understanding the fundamental knowledge of materials for any sustainable applications. Definition and concept of renewable and sustainable materials, green technology with a brief history behind each term, are presented.

In the eyes of many, the words 'chemistry' or 'chemical' can never be put on the same page with 'sustainable', 'sustainability' or 'green technology'. Despite the negative connotations associated with the word 'chemistry', in reality many chemical principles and processes are used to develop renewable and sustainable materials. In order to do so, a wide selection of organic materials is readily available. A detailed discussion on chemistry and chemical approach of these materials is presented in Chap. 2. We hope in the near future, chemistry for sustainability would become the new mantra in the world of ever-limited resources.

Materials are very important in the development of industries across the world. Production of materials can be developed either through physical process, chemical process or biological process. Looking at the angle of biological term, this process involves living organisms but is made up of many chemical reactions or other mechanisms that are related to persistence and transformation of life forms. This process plays significant role in many industries such as medical, cosmetics, food, agriculture, and so on as one of sustainable and green technological options. With the objective to provide an insight into this, Chap. 3 presents fundamental principles of renewable and sustainable material production based on biological approach including bio-based techniques of downstream and upstream processing.

In the recent years, the environmental issues in terms of pollution and energy consumption foment an increasing interest to focus on green technology. Pertaining this, Chap. 4 is intended to recapitulate a review on renewable and sustainable materials produced either from single unit process or integrated approach through chemical or biological methods or combination of these methods for various green technology applications. This includes the recent development and potential of these materials for energy, building, and environmental applications.

Every effort towards renewability and sustainability is not without its hurdles. Factors such as economy, technical, and management are the biggest challenge that must be overcome. In a nutshell, it requires the collective effort from everyone; legislative, scientists, industries, and the rest of people to make it happen. With this regard, Chap. 5 outlines the challenges towards sustainability. After water and food security, the next concern is the renewable and sustainable energy. Since the fossil fuel is still the main thrust in the current economy, the alternatives should already be in the pipeline by now. Harvesting natural resources for the fossil fuels replacement must be done without bringing negative impacts to the ecosystem. Despite the tough current situation, the future of the world still looks bright.

In addition, the work involved in this book is part of the outcomes of funded research projects, and thus we would like to take this opportunity to express our sincere appreciation to the sponsors for the financial and technical supports. Our thanks go to MOHE FRGS Grant 203/PTEKIND/ 6711574, USM Short Term Grant (304/PTEKIND/6315021), and MOHE FRGS Grant (203/PTEKIND/6711373). We also appreciate our graduate research students for their assistance in performing and conducting excellent research works associated with these funded projects. Our appreciation also goes to School of Industrial Technology, Universiti Sains Malaysia for facilitating the process of gathering material and information necessary for publishing this book. Our special thanks to Anthony Doyle, Vani Gopi, Chandra Sekaran Arjunan, Suganya Manoharan, and the editorial team of Springer International Publishing for their contribution in any kind of forms in bringing the book to fruition. In addition, we also appreciate all reviewers for their time reviewing the content of this book. Last but not least, we hope that this book would serve as a reference to academics, researchers, professionals, and students in working in this field. This book is our labour of love, may you enjoy reading it as much as we enjoyed writing it.

Penang, Malaysia
December 2017

Mohd Firdaus Yhaya
Husnul Azan Tajarudin
Mardiana Idayu Ahmad

Contents

About the Authors

Mohd Firdaus Yhaya obtained his Ph.D. from the University of New South Wales, Australia in 2012. During his stint there, he worked under the supervision of Professor Dr. Martina Heide Stenzel, a renowned polymer chemist. He is currently a Senior Lecturer in Division of Bioresource, Paper, and Coatings (BPC) Technology, School of Industrial Technology, Universiti Sains Malaysia. His research interests are in angstrom technology, polymer chemistry, organic chemistry, coatings chemistry and technology, click chemistry, and green chemistry. He is focusing on patenting novel processes and chemistries that have greater impact on the modern societies. Currently, he is working on sulphurless curing system (neither radiation nor peroxide curing) for natural rubber latex and freeze-resistant natural rubber latex with long storage capability with Associate Professor Dr. Baharin Azahari and Associate Professor Dr. Azura A. Rashid. Teamed with Associate Professor Dr. Mardiana Idayu Ahmad, they are working on a module that is capable of reducing the air-conditioner workload, thus reducing the electricity bills dramatically. He is also working together with Dr. Husnul Azan Tajarudin on recovery of valuable fatty acids from leachate fermentation. Together with Dr. Hayati Samsudin, both of them are working on polymeric packaging that is able to extend the food's shelf-life. He is also cooperating with Professor Dr. Rokiah Hashim on development of novel flame retardant particleboards that may save lives. Regardless of what the *naysayers* think, he is always believe in the impossible.

Husnul Azan Tajarudin is currently Senior Lecturer at the Division of Bioprocess Technology, School of Industrial Technology, Universiti Sains Malaysia, Malaysia. He received his Ph.D. from Swansea University, United Kingdom in 2012 in the area of Bioprocess Engineering. He is a Certified Environmental Professional in IETS Operation Biological Process and also a Registered Environmental Engineer under Board of Engineer Malaysia. His main research areas are on environmental bioprocess technology and engineering with speciality on conversion waste to wealth and treatment system.

Mardiana Idayu Ahmad obtained her Ph.D. in Engineering Science: Sustainable Energy Technologies at the Department of Architecture and Built Environment, Faculty of Engineering, University of Nottingham, United Kingdom in 2011. She is currently Associate Professor in the Environmental Technology Division, School of Industrial Technology, Universiti Sains Malaysia. Her research spans in the breadth of energy and environmental technologies. She has always been passionate about continuing her research in a way to bridge these two fields. Her research work leads to the production of over than 100 publications nationally and internationally, including journal papers, research books, popular academic books, book chapters, conference proceedings, and other publications. With high passion for writing, she has also written short story books for children as primary target readers and this has led to the enhancement and development of knowledge to the community.

Abbreviations

3HHx	3-hydroxyhexanoate
3HV	3-hydroxyvalerate
ACF	Activated carbon fabric
CHNOPS	Carbon, Hydrogen, Nitrogen, Oxygen, Phosphorus, and Sulphur
CMB	Construction microbial biotechnology
CoA	CoA
CS_2	Carbon disulphide
CSTR	Continuous stirred tank reactor
DNA	Deoxyribonucleic acid
EDTA	Ethylene diamine tetraacetic acid
GAP	Glyceraldehyde-3-phosphate
GlpD	Glycerol-3-phosphate dehydrogenase
GM	Genetically modified
GMO	Genetically modified organism
HCl	Hydrogen chloride
HCN	Hydrogen cyanide
HDPE	High-density polyethylene
LNG	Liquefied natural gas
MBR	Membrane bioreactor
MCL	Medium chain length
MCP	Microbial carbonate precipitation
MICCP	Microbially induced calcium carbonate precipitation
NaCl	Sodium chloride
NaOCl	Sodium hypochlorite
P(3HB-3HV)	Poly(3-hydroxybutyrate-co-3-hydroxyvalerate)
P3HB, PHB	Poly 3-hydroxybutyrate
P4HB	Poly-4-hydroxybutyrate
PAN	Polyacrylonitrile
PBS	Polybutylene succinate

PCBs	Polychlorinated biphenyls
PCL	Polycaprolactone
PDO, 1,3-PD	1,3-Propanediol
PET	Polyethylene terephthalate
PGA	Polyglycolic acid
PHA	Polyhydroxyalkanoates
PhaA	Acetyl-CoA by b-ketothiolase
PhaB	Acetoacetyl-CoA reductase
PhaC	PHA synthase
PHBHHx	Copolymers of 3-hydroxybutyrate and 3-hydroxyhexanoate
PHBV, PHB-co-PHV	Copolymers of 3-hydroxybutyrate and 3-hydroxyvalerate
PHO	Poly 3-hydroxyoctanoate
PHV	Poly-3-hydroxyvalerate
PLA	Polyhydroxyalkanoates
PM	Particulate matters
POME	Palm oil mill effluent
PP	Polypropylene
PS	Polystyrene
PVC	Polyvinyl chloride
RNA	Ribonucleic acid
ROI	Return of investment
RSC	Royal Society of Chemistry
SCL	Short chain length
TPI	Triosephosphate isomerase
UN	United Nations
UNFCCC	United Nations Framework Convention on Climate Change
USA	The United States of America
USAB	Upward-flow anaerobic sludge blanket

Chapter 1
Introduction

1.1 Background

Materials are important in human civilisation stages started from Stone Age, throughout Copper Age, Bronze Age, and to the Iron Age, but what are they, really? By definition, materials are substances of which things are made or composed. In simple technical language, materials are atoms combined together in the solid state. Studies of materials started when early researchers in the Age of Enlightenment began to utilise analytical thinking from chemistry, physics, and engineering to apprehend other fields such as metallurgy, mineralogy, ancient, and phenomenological observations which later sparked out the field of material science (Smith 1981). With the turn of the century, material science is becoming one of very crucial aspects in science stream involving fundamental physical and chemical basis of the combination atoms or atomic bonding to form new compounds, phases and structures in order to meet the new era demands. This revolution of material science leads to innovations and sustainable solutions to technological societal and environmental issues.

Selection of materials is essential in process or development either in industrial designs or engineering applications. Failures to choose good or best materials will result in unpredictable impact for long-term success of any applications. In the selection of materials, for instance, for an application that requires a high tensile strength, a material with good and higher tensile strength must be chosen. Also, if a product is intended to be produced for outdoor usage, it may be necessary to consider the effects of outdoor conditions. So, a systematic approach is needed in selecting the best materials for a particular application which requires a proper technique. In this technique, application requirements in terms of mechanical, thermal, environmental, physical and/or biological properties should be considered. This chapter presents a review of a few important keywords within the scope of this book with the aim to provide a foundation for understanding fundamental knowledge of materials for any sustainable applications.

M. F. Yhaya et al., *Renewable and Sustainable Materials in Green Technology*, SpringerBriefs in Applied Sciences and Technology,
https://doi.org/10.1007/978-3-319-75121-4_1

1.2 Renewable and Sustainable Material

A renewable or sustainable material is also named as a sustainable resource which can be generated or produced and supported by nature. It can be renewed without depleting non-renewable resources or without disturbing the established equilibrium of natural environment and friendly with the ecosystem. Renewable or sustainable materials can be manufactured or generated quickly enough to keep pace with how fast they are used up as compared to non-renewable materials, which take a long time to renew. Besides, they can also be produced in high enough volume to be economically useful. Therefore, in general, we can say that these materials are natural, renewable, carbon neutral, use less energy to extract, non-toxic or harmless and economical to produce. They can be made from either natural, synthetic, or recyled products.

Therefore, we should comprehend the fundamental science behind new forms of renewable and sustainable materials and, once promising opportunities are identified, new materials that meet the demand of industry and commerce, reduce adverse environmental impacts, and decrease sociological effects of conventional materials can be generated. In understanding this, there are a few fundamental questions to ask as a starting point to assess the sustainability of any materials. The questions are as follows:

- What is its real value—for initial use and long term?
- Does it provide optimal performance for its application?
- Is it widely available?
- How ubiquitous are the source materials? Or, do they regenerate and how quickly?
- What is needed to process it into a usable form? Did this process produce or release toxins or destroy habitat?
- How much energy and water did it take to make it?
- How much waste material did it generate?
- What does it need to operate—maintenance inputs, operating energy?
- Were the people involved in producing, delivering and installing it fairly compensated? Were they provided with safe and healthy working conditions?
- How long will it last? What happens at the end of its service life?

In order to have a better understanding of these questions, the meaning of sustainability should be looked into. Sustainability means to 'meet the needs of the present without compromising the ability of future generations to meet their own needs' (WCED 1987). Sustainability also relates to the dynamic condition of the earth's biosphere and its various systems with the harmony and balance between human and nonhuman elements (Flint 2013). So, in order to make a decision involving sustainability aspect, it entails scientifically based information on sustainability (Dong and Hauschild 2017). In short, sustainability of materials looks to safeguard our natural environment, human and ecological health, while driving innovations and technologies without compromising our way of life.

In addition, in selecting renewable and sustainable materials, dealing with sustainability issue, a concept of 'cradle to grave' should be applied and assessed (Giwa 2017; Tsang et al. 2016). This concept is important in the production of materials in relation to the effect of the materials or products to the environment. It involves the lifecycle of a material or product from its creation to its disposal. In this part, life cycle assessment plays an important role to guide the selection of materials for sustainability option.

1.3 Green Technology

The words 'green technology' consists of two different words that have a major impact to the world these days. It is worth to get a basic overview of the definition of these two words before we dig more into this field. The term 'technology' refers to the branch or application of scientific knowledge for practical purposes, which deals with the creation and usage of technical means in engineering, applied sciences, pure sciences, industrial arts and their interconnection with life, society and the environment. 'Green' is a common word which indicates colour of nature (plants) in general, and nowadays it has become a popular term which represents something positive and beneficial to the environment or less harmful to the environment than others. So, the combination of these two words 'green technology' refers to a continuously evolving group of materials and methods from scientific techniques or knowledge to generate energy-efficient to non-toxic products. In other words, this technology encompasses innovation or invention that has features, functionality and cost savings elements which would give advantages to the environment and society. It is also called as clean technology or environmentally friendly technology.

People usually think that this concept is new and has just started to drive attention in the twentieth century. The truth is, the concept of green technology is not new and has been around for centuries. The earliest known application of green technology in the form of passive energy source from solar arose during the fifteenth century B.C. by the Egyptian ruler Amenhotep III with regard to the operation of his 'sounding statues' (Barber 2017). In the fifth century A.D., Socrates noticed that buildings should face towards the south so that the sun could penetrate them during winter and provide light and warmth (Butti and Perlin 1980). Solar architecture was then employed by the Greeks and Chinese about two and a half millennia.

In this new era, in the attempts to pave the way for a low-carbon future as a feature of modern times, people nowadays took more seriously on green technologies. With the target to boosting energy from renewable sources, together with combating climate change and environmental issues, green technologies have now become the main component of governments' agenda of most countries around the globe. As part of this, investment is growing in green technologies in order to fulfil the demand of green economy (Cao and Wang 2017). In this regard, most of the countries have now rapidly shifted towards sustainable solutions with burgeoning innovations that have a clear environmental and social attention including renewable and energy-efficient technologies, water purification, carbon dioxide (CO_2) conversion, nanoscale bio-based engineering, passive buildings, and molecular nutrition. As green technologies continue to arise as a growing dynamism, several strong industry clusters have emerged such as water and wastewater, advanced materials, energy, agriculture, transportation, waste management, buildings, and manufacturing.

References

S. Barber, History of passive solar history, East Carolina University. http://studylib.net/doc/12033327/history-of-passive-solar-energy-scott-barber. Accessed on 1 Nov 2017. (2017)

K. Butti, J. Perlin, *A golden thread: 2500 years of solar architecture and technology* (Cheshire Books, USA, 1980)

B. Cao, S. Wang, Opening up, international trade, and green technology progress. J. Cleaner Prod. **142**(2), 1002–1012 (2017)

Y. Dong, M.Z. Hauschild, Indicators for environmental sustainability. Procedia CIRP **61**, 697–702 (2017)

W. Flint, *Practice of sustainable community development: a participatory framework for change* (Springer International Publishing, New York, USA, 2013)

A. Giwa, Comparative cradle-to-grave life cycle assessment of biogas production from marine algae and cattle manure biorefineries. Bioresour. Technol. **244**(2), 1470–1479 (2017)

C.S. Smith, *A search for structure* (MIT Press, USA, 1981)

M.P. Tsang, G.W. Sonnemann, D.M. Bassani, Life-cycle assessment of cradle-to-grave opportunities and environmental impacts of organic photovoltaic solar panels compared to conventional technologies. Solar Energy Mater. Solar Cells. **156**, 37–48 (2016)

WCED, *Our common future. World commission on environment and development* (Oxford University Press, Oxford, UK, 1987)

Chapter 2
Renewable and Sustainable Materials from Chemical Approach

2.1 Chemistry Is the Central of Science

Chemistry was first coined as the central of science in 1976 because it interconnected and formed the base of physical sciences, life sciences and applied sciences together (Brown and Lemay 1976). Physical sciences are the likes of physics, astronomy, chemistry and earth science. Life sciences are dedicated to the study of living organisms. Applied sciences are putting all those proven science theories to good use, just like medical sciences and engineering. Starting from the formation of atoms, elements and molecules, chemistry can explain many phenomena observed in physical sciences, life sciences and applied sciences.

For example, consider nylon. It is a synthetic polymer that can be turned into many useful materials; among them are toothbrush bristles and stockings. In terms of physical properties, it is a strong polymer and can be explained from chemistry viewpoint. The strength is due to the polymer chains that able to form hydrogen bonding between them. Ants sting in order to protect themselves and their colonies. The sting contains formic acid (methanoic acid) which belongs to the carboxylic acid family. In civil engineering, buildings in the earthquake-prone areas are built on natural rubber-based foundation that will absorb the violent lateral forces, saving them from collapsing. From chemistry viewpoint, this is due to the cross-linking of the natural rubber chains that provides the elasticity.

However, in the eyes of many, 'chemistry' or 'chemical' can never be put in the same sentence with 'sustainability'. Any mention of the word 'chemistry', the things that come to mind are symbol of skull with crossbones, pollution, dumping of toxic waste and others. According to survey by Royal Society of Chemistry (RSC) in 2015, people are being aware of chemistry's contribution to society and they tend not to hold very strong negative views about it. However, it should be taken into consideration that this survey was done in United Kingdom. The situation may differ if the survey was done in other countries where the public awareness is still limited. Despite this misunderstanding, chemistry principles are still used for developing renewable and sustainable materials now and in the future.

M. F. Yhaya et al., *Renewable and Sustainable Materials in Green Technology*, SpringerBriefs in Applied Sciences and Technology,
https://doi.org/10.1007/978-3-319-75121-4_2

2.2 Organic Materials for Sustainability

Organic, inorganic or combinations of both (hybrid) materials can be used as renewable and sustainable materials. Inorganic and hybrid materials are already well discussed in the literature, so this chapter will discuss more on organic materials. In terms of progression, more attention is directed towards organic materials due to their abundance and ease of processing as compared to inorganic and hybrid materials.

Organic chemistry means chemistry of carbon compounds, so organic materials are made of carbon compounds. Living organisms are built upon these six elements: CHNOPS (carbon, hydrogen, nitrogen, oxygen, phosphorus and sulphur). Until recently, there is no official definition for organic materials. In order to be considered as organic materials, usually these three requirements should be observed. First, the material must be a carbon compound. Second, it must be found in or originated from living things. Third, the carbon must be attached directly to hydrogen atom/s by covalent bonding. Petroleum, natural gas, coal and peat are organic because they were once living organisms. Pure carbon element, diamond, graphite, buckminsterfullerene and calcium carbonate are not considered as organic materials. However, some exceptions do exist. Carbon monoxide and carbon dioxide are considered as inorganic even though they contain carbon atoms and/or originated from living organisms. Urea and oxalic acid are organic because they can be obtained from living things despite none of their carbon atoms are attached directly to the hydrogen atoms. All organic compounds must contain carbon, but some carbon-containing compounds are not organic.

2.2.1 *Lignocellulosics Biomasses*

Lignocellulose is usually obtained from plant dry biomass. It is composed of aromatic polymer (lignin) and carbohydrate polymers (cellulose and hemicellulose), and hence the name lignocellulose. Lignocellulose resources are abundant and also renewable. From the early human history, the main source of lignocellulosic materials is wood. Wood has been used for fuel (cooking and to provide heat against winter), construction, furniture and papermaking. Unfortunately, the demand for wood is a major cause of damage to forests. In order to access the trees, the jungle needs to be cleared to make way for timber trucks. This will cause loss of plant species, displacement of wild animals, soil erosion, contamination and sedimentation in rivers. As a solution, wood from sustainable plantation is used (such as for papermaking) or by utilising agricultural waste instead.

Dry lignocellulosic biomass from sugarcane bagasse, palm oil empty fruit bunches and others are grounded into fine particles and glued together to fabricate particleboard panels for furniture making. Depending on the geographical locations, different countries may use different crops to produce lignocellulosic biomass. For example, in temperate countries, corn, wheat or barley may be used while sugarcane, coconut,

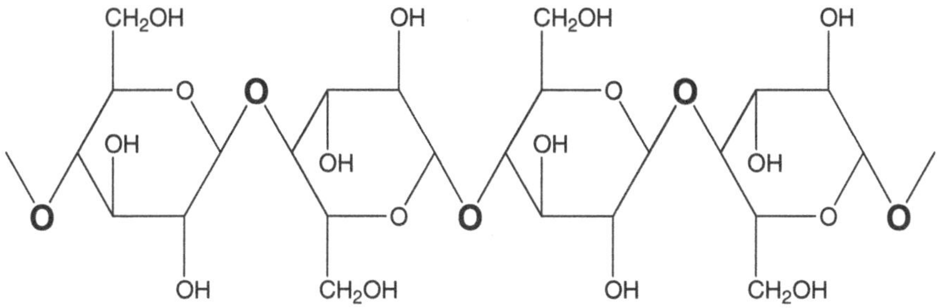

Fig. 2.1 Chemical structure of cellulose. Note that the oxygen atoms highlighted at the joining points are alternatingly pointing upwards and downwards the plane

kenaf, banana and oil palm waste may be used in tropical countries. Previously after harvesting, this lignocellulosic biomass is dumped, used as fertiliser, animal feed, or burn to generate energy. Due to the environmental awareness, these biomasses have been studied to produce paper, biocomposites, furniture, insulation, packaging and so on.

The most important component in lignocellulosic biomass is cellulose. Cellulose can be classified as a natural, condensation and carbohydrate polymer. It is a linear polymer made from glucose with strong intermolecular hydrogen bonding that is difficult to penetrate and replace. That is the reason cellulose does not dissolve in water despite the abundance of hydroxyl groups (Fig. 2.1).

For the same reason, cellulose cannot be dissolved in organic solvents directly. The purpose of dissolving the cellulose is to provide accessibility to the most of the hydroxyl groups. The making of rayon and cellophane requires the replacement of cellulose hydroxyl groups with other functional groups such as acetate or xanthate, later formed into film or fibre, before regeneration back into cellulose. Although rayon and cellophane are based on cellulose which is natural, abundant and biodegradable, their processing is water- and energy-intensive. The usage of toxic sodium hypochlorite (NaOCl), tetraamine copper dihydroxide and carbon disulphide (CS_2) has contributed to the air and water pollution. Another less polluting way of dissolving cellulose is using ionic liquids. Ionic liquids are salt. They exist in liquid form because the ions cannot be packed closely together like sodium chloride (NaCl) due to steric hindrance. These liquids can be recovered at the end of the processing. At the moment, the cost of ionic liquids is still high that may limit its usage in the industry.

Chemical modifications can still be done without fully dissolving the cellulose. The downside is that not all of the hydroxyl groups are able to take part. In the production of particleboard for furniture, wood/biomass particles are glued together with phenol-formaldehyde resin using heat and pressure. Some of the hydroxyl groups on the surface of the particles are involved during the curing process. Even though not all hydroxyls groups are involved, it is just enough for the particles to be joined (cross-linked) together to form a board. The paper made from cross-linked empty fruit bunch fibres was found to have greater wet tensile strength than the uncross-linked ones (Rahman et al. 2016).

Lignin is another component in lignocellulosic material (Fig. 2.2). It is amorphous, has lower biodegradability relative to cellulose and starch, covalently bonded to the cellulose, and reinforces the cell wall and plant as a whole. Wood is brown colour due to the existence of lignin. The aromatic rings in its chemical structure are responsible for its strength and ultraviolet light absorbance. It is made of randomly cross-linked phenylpropane units such as coumaryl alcohol, coniferyl alcohol and syringyl (sinapyl) alcohol. These alcohol units were found on the smoked food such as barbeque due to the thermal degradation of lignin. The chemical structure is complex, due to the many possible bonding patterns between individual units. Lignin is usually separated from the cellulose via pulping process as part of the concentrated black liquor. It is usually burned as fuel for energy to run the mill. At the moment, there is strong interest to recover lignin from pulping at large scale (Lake and Blackburn 2014). The recovered lignin may be used as aromatic feedstock for industrial use. The lignin potential should not be underestimated since it is the second most abundant organic compound after cellulose. Global research work so far has managed to fabricate polyurethane foam, carbon fibre, activated carbon and mouldable plastic out of lignin.

2.2.2 Edible and Inedible Starches

Edible starches are safe to be consumed by human and animals, originated from rice, flour, corn, potatoes, cassava, sago and many more. Although tapioca starch and rubber seeds contain linamarin that will liberate toxic hydrogen cyanide, this problem can be overcome by soaking in running water (river) for a few days, charcoal grilling and cooking. Chemical modification of edible starches turns them into resistant starches because they could not be recognised by the human enzymes. Since these starches are not broken down into glucose, they should be good for diabetic patients, for those watching their body weight and for improving bowel movements.

Inedible starches contain toxic, including elephant cassava/tree cassava/Ceara rubber tree (*Manihot glaziovii*), intoxicating yam (*Dioscorea daemona*), *horse* chestnuts (*Aesculus* spp.) and Osage orange (*Maclura pomifera*). However, latex from elephant cassava may become alternative source of natural rubber, currently dominated by rubber trees (*Hevea brasiliensis*). These starches are more suitable to be fermented in order to produce ethanol, provided the selection of microorganisms and acclimatisation is done properly (Moshi et al. 2015).

Starch consists of linear, helical amylose and the branched amylopectin (Fig. 2.3). In many ways, starch is similar to cellulose as far as chemical properties are concerned. However, the similarities end there. The different arrangements of glucose monomers in cellulose and starch determines their physical properties. Glucose dissolves in both cold and hot water. Starch dissolves only in hot water, while cellulose does not dissolve in any one of them. As compared to the cellulose, the glucose

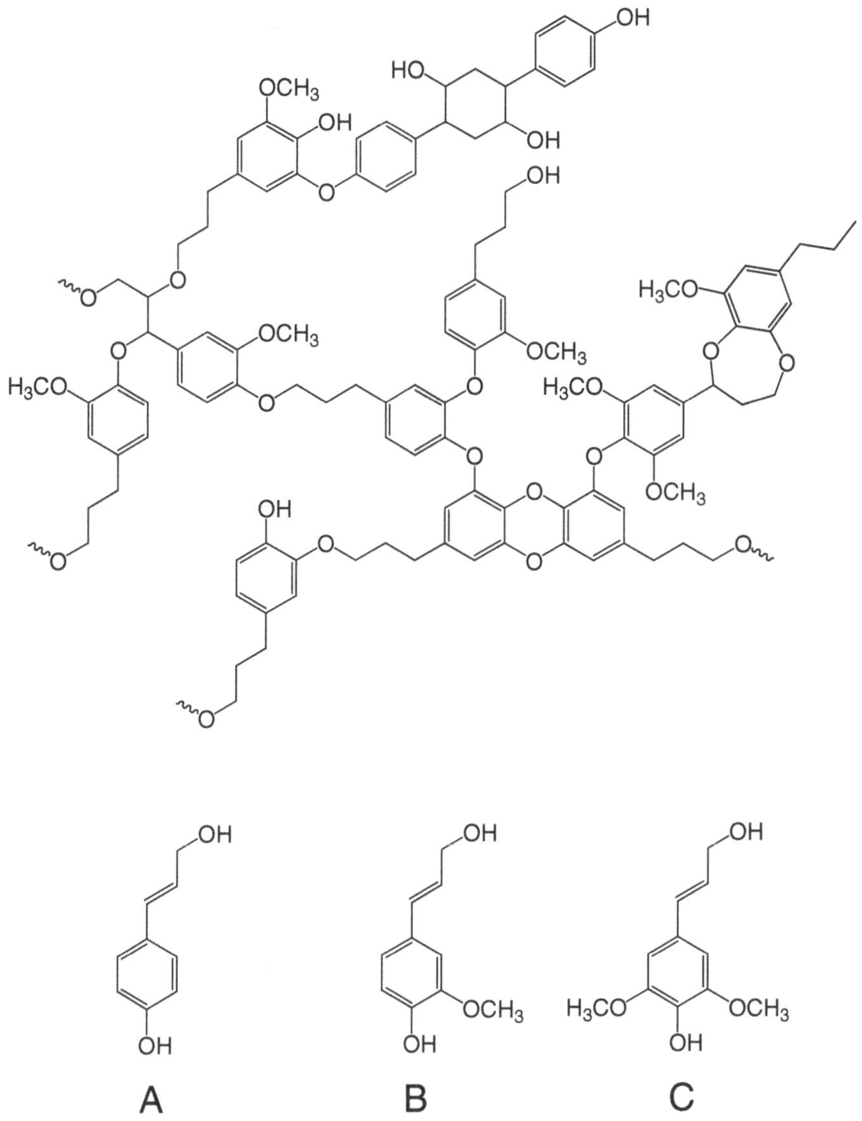

Fig. 2.2 Chemical structure for varieties of lignin. Phenylpropane monomers making up the lignin including coumaryl alcohol (**a**), coniferyl alcohol (**b**) and syringyl alcohol (**c**)

monomers are arranged with the oxygen atoms at 1, 4-linkages that are all pointing upwards. Starch is therefore easier to modify chemically as compared to cellulose. Chemical modifications usually done to starch are acetylation, ionification and cross-linking. The secret lies in the hydroxyl groups (–OH) at carbon number 2, 3 and 6

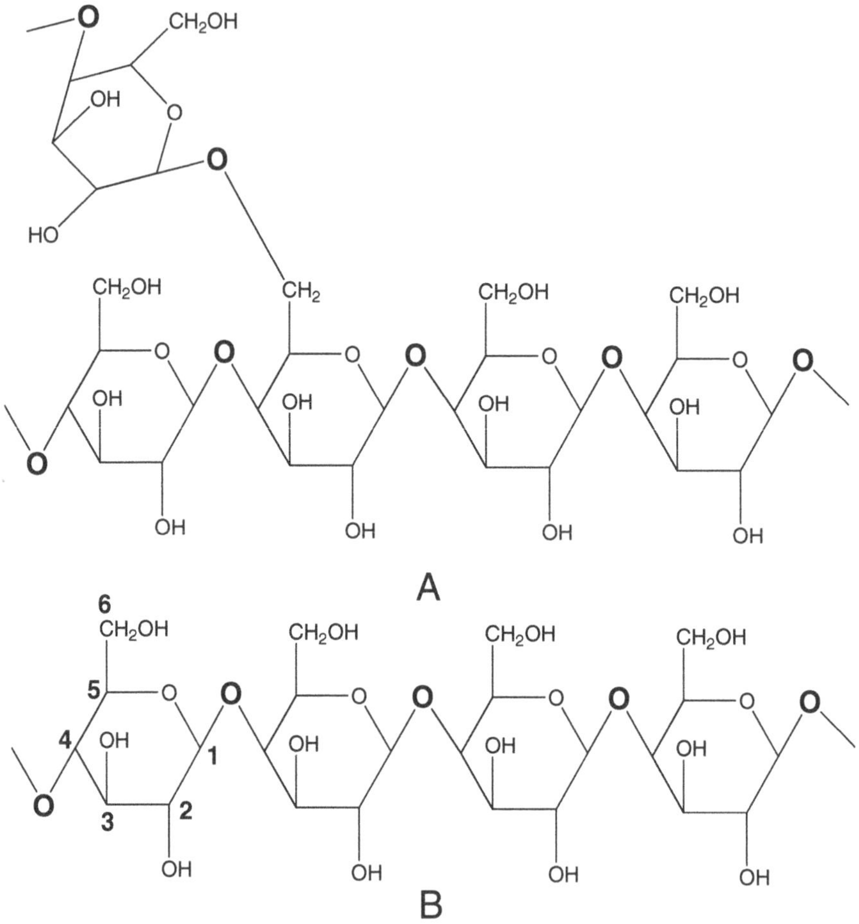

Fig. 2.3 Chemical structure of starch. Note that the oxygen atoms highlighted at the joining points are all pointing upwards the plane. **a** is the branched amylopectin, while **b** is the linear amylose. The numbers from 1 to 6 refer to the carbon numbering

(Fig. 2.3) that are easily modified. In cellulose, the modifications are much more difficult since some hydroxyl groups are inaccessible due to strong hydrogen bonding.

Amylose and amylopectin chains are contained in the starch granules. In order to chemically modify these carbohydrate polymers, the granules must first be broken down by heat and water through a process called starch gelatinization. This process will break down the helix structure of amylose chains, releasing the chains out of the granules. Meanwhile, due to branching, amylopectin chains are stuck and left in the granules. Towards the end, the granules containing amylopectin are held in the matrix of amylose, forming a gel (network) that traps the water molecules inside. This is the reason for the increased viscosity of starch paste upon cooking. In daily life, this process can be observed by cooking rice in a pot. During gelatinization, the lid of the pot is pushed upwards and the content inside is bubbling and spills onto the side. If this process does not happen, the rice is considered uncooked or undercooked. If eaten, it may cause stomach ache to some people due to the problem with digestion.

Due to the fact that starches are renewable, biodegradable and relatively cheap, they have been used in the production of plastic packaging. Starches may be added or grafted into polyethylene backbone in which the former will degrade after disposal. Some companies are making packaging with almost 100% of starch content. In the environment, starch molecules are hydrolysed by fungi and bacteria into glucose for consumption and finally turned into water and carbon dioxide.

2.2.3 Vegetable and Animal Oils

Natural and edible oils from vegetables and animals are renewable and also sustainable. These oils are mainly based on triglycerides (Fig. 2.4), a natural combination of glycerol with three fatty acids. Vegetable oils can be obtained from seeds/fruits of corn, rapeseed, peanut, olive, soya, coconut, palm oil and others. Fish oils are the most popular animal oils despite the availability of other sources. Vegetable oils are one of the components in making alkyd resin for coatings and printing inks. Physically, the difference between oils and fats is that oils are liquid while fats are solid at room temperature. Chemically, oils contain more monounsaturated and polyunsaturated fatty acids while fats contain more saturated and/or trans fatty acids. Fatty acids are carboxylic acids with long hydrocarbon chains (Fig. 2.4). Saturated means no –C = C– double bonds. Different sources of triglycerides have different compositions and percentages of fatty acids. Triglycerides may be broken down into glycerol and various fatty acids by hydrolysis with strong base (sodium hydroxide). Glycerol (glycerin/glycerine) has been used in food industry as additive and moisturiser. It is colourless, odourless, non-toxic and a polyol (triol) that is polar and water soluble. It is the monomer for many condensation (step-growth) polymerisations. Fatty acids are carboxylic acids with the hydrophilic –COOH part and the hydrophobic alkyl part.

2.2.4 Recycled Plastics

Recycled plastics may also be used for sustainability, whether we believe it or not. Everywhere, common plastics are regarded as nuisance since they are not degradable. Most of these culprits are used for packaging materials. If disposed improperly, they can block the drain and sewage system, causing flash floods. Once entering river and finally end up to the sea, they mislead turtles into thinking them as jellyfish, end up

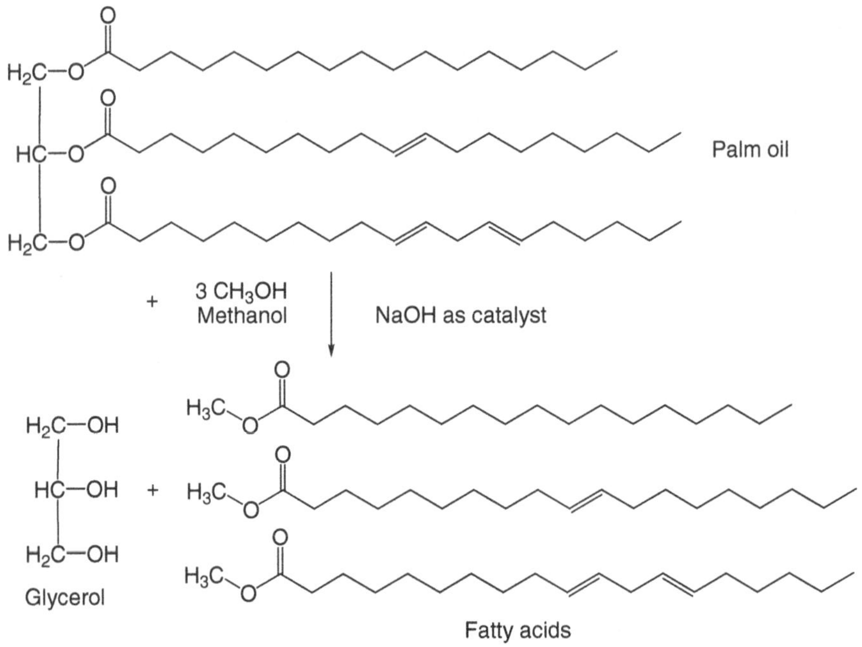

Palm oil

+ 3 CH_3OH
Methanol

NaOH as catalyst

Glycerol

Fatty acids

Fig. 2.4 Transesterification of palm oil triglycerides into glycerol and fatty acids

choking and finally killing them. Plastic bottles have also reported to be found in the stomach of a dead whale lying on the beach.

Plastics in general can be divided into thermoplastics and thermosets. They are recognised by the recycling triangle imprinted on the products. Thermoplastics can be melted and dissolved in organic solvents many times while thermosetting plastics cannot. Both can be recycled using proper techniques. Polyethylene terephthalate (PET) is thermoplastic polyester normally used for making soft drink bottles and fibres. After disposal, PET bottles can be collected, cleaned up and turned back into its original precursors (ethylene glycol and terephthalic acid). The Tzu Chi Foundation in Taiwan, for instance, is currently turning used PET bottles into blankets for disaster relief. In rural areas of Bangladesh, where electricity does not exist, the PET soft drink bottles are cut into half and used as the world's first zero-electricity air cooler. This will reduce the temperature by up to 5 °C from the original 45 °C during the summer. High-density polyethylene (HDPE) can be recycled into recycling bins, ropes and pipes. Polypropylene (PP) is a tough plastic and can be recycled up to four times before the effect of thermal degradation takes place. The physical strength keeps on reducing as PP undergoes new recycling process. Polystyrene (PS) is much difficult to recycle since it cannot be recycled locally and must be sent to the recycling centre. Transportation of light yet bulky PS is uneconomical. The least can be done with PS to turn it into composites.

Thermosets are even more difficult to be recycled since their polymeric chains are tightly joined together by covalent bonding and cannot flow upon heating. Heating will only cause the thermosetting plastics to decompose. Despite this limitation, recycling can still be done. The thermoset can be grounded into smaller particles and used as filler in the making of new products. Used automobile tyres made of cross-linked rubbers are shredded, removed of their steel carcasses and finally used as a raw material for civil engineering works (road pavement, collision reduction and rainwater runoff) and floor covering for playgrounds. Pyrolysis of used tyres produces valuable steel, oil, char and gas (Williams 2013). Oils from pyrolysis are composed of aliphatic, aromatic and polyaromatic hydrocarbons. Polyurethane foams are grounded into small particles for carpet backing. So far, the progress in recycling of cross-linked epoxies is very slow. They are usually burned as fuel for the energy.

2.2.5 *Sewage Sludge*

Sewage is a semi-solid residual byproduct of sewage treatment of municipal and industrial wastewater. Sewage sludge may contain combination of pathogens, heavy metals, antibiotics and organic materials. The final place for sewage sludge includes incineration, landfill and used as fertiliser. There are some issues with each of the methods described. Incineration may produce toxic gases containing dioxin, polychlorinated biphenyls (PCBs) and nanoparticles that required proper monitoring and maintenance to prevent them from leaking into the environment. Disposal of the remaining ash is also a concern. Landfilling of sewage sludge may pose health risk due to bioaerosols, pathogens and endocrine disruptor chemicals such as triclosan and triclocarban. The usage of sewage sludge as fertiliser can also pose health risk due to the contamination of farmland with heavy metals. Heavy metals and endocrine disruptor may accumulate in plants and transferred to human upon consumption. The best practice at the moment is to restrict the usage of toxic materials so that they do not enter sewage system at the first place and reducing sludge production using dry toilets instead of flush ones.

Another problem is the palm oil mill effluent (POME) originating from oil palm milling. Currently, the world's leading vegetable oil in terms of production volume and consumption is palm oil. High volume production leads to high effluent volume. The effluent contains mostly water with residual oil and suspended solids. The problem with POME is that it is high in biochemical demand, low pH, produces huge amount of greenhouse gases (methane and carbon dioxide), stinky, and will contaminate the waterways if overflowed during the monsoon seasons. Much has already been done since yesteryears to improve this situation. The POME has been treated using conventional biological treatment systems such as anaerobic, aerobic and facultative ponds in series. POME was found to be better used for biogas/methane production rather than for livestock feeding, as a 40 tonnes/hectare oil mill would require around 44,000 pigs or 43,000 cattle for the entire effluent to be utilised (Agamuthu 2016).

2.2.6 Manures

Manures are organic matter derived from animal or human faeces which have been long used as plant fertilisers. There are three issues related to the usage of manures. The first is that manure may be a source of pathogens carried by vector organisms which may cause disease or risking food safety. The second issue is that antibiotics have been found in vegetables planted using fertiliser made of animal manure. Plants have the ability to absorb and accumulate these antibiotics inside them. Once consumed, these antibiotics may enter human bodies, causing antibiotics resistance. The last one is the smell produced by manure. Pig manure may cause the most unforgivable stenches to those residing close to the pig breeding farm, followed by other manures.

2.2.7 Biopolymers and Biodegradable Polymers

The problem with synthetic polymers is that most of them are based on petroleum which is non-renewable. Even worse, most of synthetic polymers are also non-biodegradable with few exceptions. In order to solve the problem, biopolymers and biodegradable polymers are introduced to the market. Even if these two terms are used interchangeably, they are actually different. Generally, biopolymers are usually biodegradable. Biopolymers mean polymers that are naturally occurring, synthesised by living things, or made from raw materials originating from living things. However, they are not necessarily biodegradable, for example, alkyd resin. Even though alkyd resin is partly made from vegetable oil, it does not degradable. Meanwhile, biodegradable polymers can be decomposed by living organisms, although they are not necessarily made from naturally occurring materials. Polycaprolactone (PCL) is made from chemicals derived from petroleum, yet it can be degraded by bacteria.

Examples of biopolymers are cellulose, lignin, starch, protein, polylactic acid (PLA), polyhydroxyalkanoates (PHA), deoxyribonucleic acid (DNA), ribonucleic acid (RNA) and alkyd resin (Fig. 2.5). Polycaprolactone (PCL), polybutylene succinate (PBS), polyglycolic acid (PGA) and polyvinyl alcohol (PVOH) are biodegradable polymers (Fig. 2.6).

However, it should be taken into consideration that in the landfills, the condition is anaerobic. The supply of moisture and oxygen is very limited to trigger biodegradation by the microorganisms and fungi. Even newsprint is still legible after forty decades in the landfill's anaerobic condition. In reality, degradation of biopolymers and biodegradable polymers is actually a problem itself. If the polymers are totally degraded under aerobic condition, it will be turned into carbon dioxide and water. If for some reasons, the anaerobic degradation can happen, it will produce methane gas which is even worse than carbon dioxide as far as greenhouse gases are concerned. Polymers for packaging will somehow end up in the ocean. Some will remain intact,

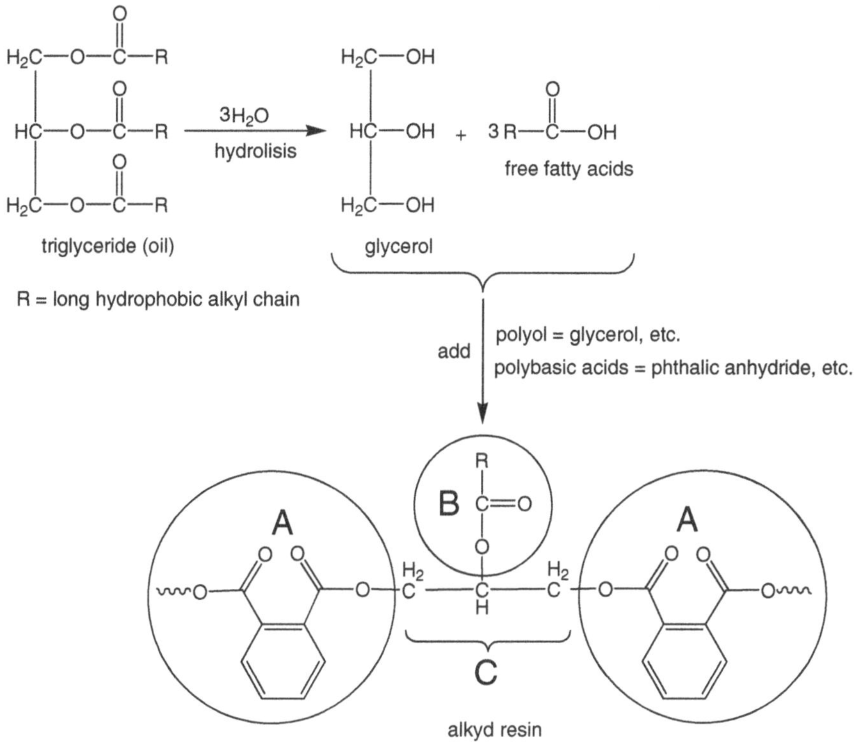

Fig. 2.5 Synthesis of alkyd resin. Even though alkyd resin is partially made from vegetable oil (triglyceride), the resulting polymer is not biodegradable. **a** = phthalic anhydride moieties, **b** = fatty acid moiety and **c** = glycerol moiety

while others are broken into smaller fragments called microplastics. These microplastics will find their way into the sea creatures and up the food chain back to humans. However, degradation is an important process to maintain the carbon cycle.

2.2.8 Lightweight Polymer Composites

Composites are made of two or more components with significant difference in physical and chemical properties. Combination of these will create a new material with different characteristics from original components. Wood is an example of natural composite. Cellulose, hemicellulose, lignin and minerals come together as one to form a strong structure that can be used as a building material and for load-bearing applications.

In synthetic composites, provided the selection of precursors is done properly, composites can possess specific strength better than steel. In airplanes, this will save

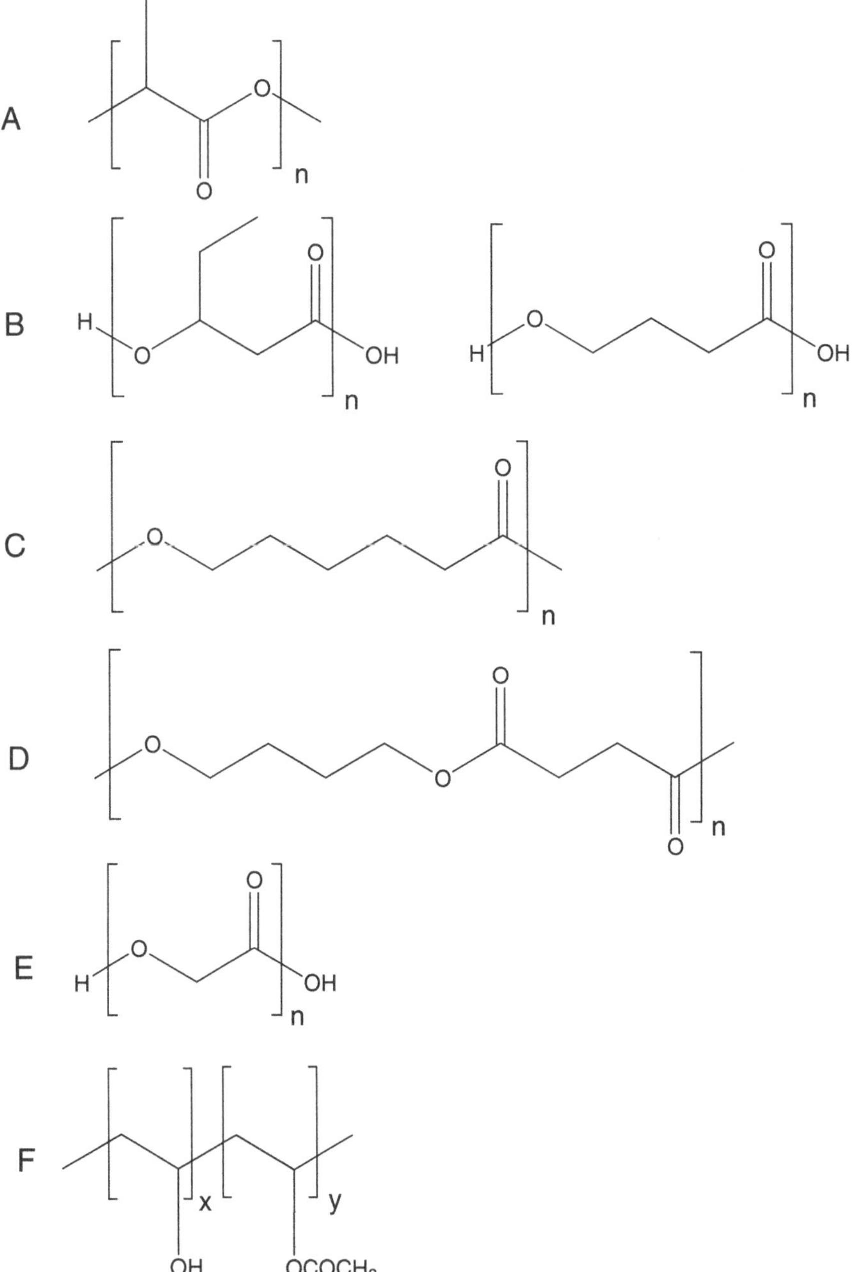

Fig. 2.6 Chemical structures for selection of some biopolymers and biodegradable polymers. **a** = polylactic acid (PLA), **b** = polyhydroxyalkanoates (PHA) based on poly-3-hydroxyvalerate (PHV) (left) and poly-4-hydroxybutyrate (P4HB) (right), **c** = polycaprolactone (PCL), **d** = polybutylene succinate (PBS), **e** = polyglycolic acid and **f** = polyvinyl alcohol. If not 100% hydrolysed, polyvinyl alcohol still contains some of the polyvinyl acetate precursor

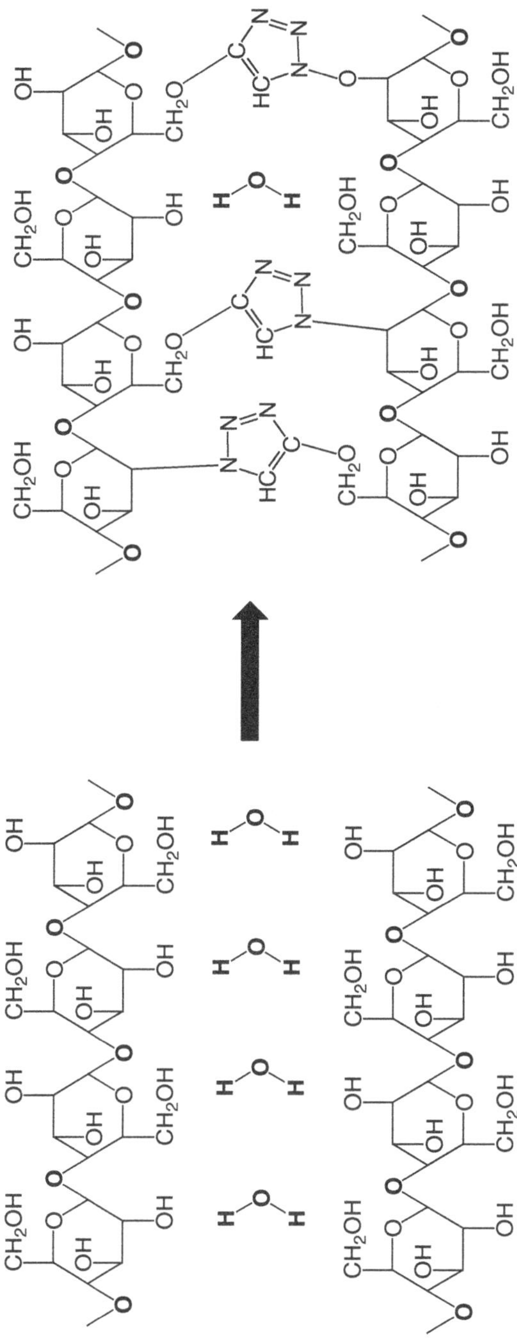

Fig. 2.7 Cross-linking of fibres through azide–alkyne click chemistry. Before cross-linking, the water molecules may cause dissociation of the fibre web easily. After cross-linking, the fibre web is linked by covalent bonding and is much stronger

the weight and reduce the amount of jet fuel used. For sports cars and supercars, extensive use of composites will boost the acceleration and top speed. These composites must be durable, long-lasting and not biodegradable due to safety requirements. For other non-critical applications, it is better to have composites that are biodegradable. Mostly, lignocellulosic fibres are mixed with biodegradable polymers to obtain biodegradable composites (Satyanarayana et al. 2009).

2.2.9 Paper with Higher Wet Strength

The basic strength of paper is made of by fibre entanglement (web) plus hydrogen bonding. In dry state, paper is quite strong, suitable for printing and packaging such as grocery bags and corrugated carton boxes. However, once in contact with water, the water molecules will find their way through, loosening up the fibres, and eventually the whole fibre web will dissociate. This problem may be solved by joining the fibres permanently by covalent bonding. Beforehand, the fibres containing cellulose and hemicelluloses are functionalized with alkyne and azide moieties. These moieties are joined together by azide–alkyne click chemistry reaction, resulting in cross-linked fibres (Fig. 2.7). The tensile strength values of the paper were found to increase in both dry and wet conditions. The increase in the latter was phenomenal, up to 4000% (Rahman et al. 2016).

References

P. Agamuthu, *Palm oil mill effluent-treatment and utilization* (Wiley, 2016), pp. 338–60

M.A. Lake, J.C. Blackburn, SLRP—an innovative lignin-recovery technology. Cellul. Chem. Technol. **48**, 799–804 (2014)

A.P. Moshi, M.M. Ken, E. Hosea, G. Elisante, G. Mamo, M. Bo, High temperature simultaneous saccharification and fermentation of starch from inedible wild cassava (*Manihot glaziovii*) to bioethanol using Caloramator boliviensi. Biores. Technol. **180**, 128–136 (2015)

N.S.A. Rahman, N.A. Ahmad, M.F. Yhaya, B. Azahari, W.R. Ismail, Crosslinking of fibers via azide-alkyne click chemistry: synthesis and characterization. J. Appl. Polym. Sci. 133 (2016)

K.G. Satyanarayana, G.C.A. Gregorio, W. Fernando, Biodegradable composites based on lignocellulosic fibers—an overview. Prog. Polym. Sci. **34**, 982–1021 (2009)

B. Theodore, H.J. Lemay, *Chemistry: the central science* (Prentice-Hall, Inc., 1976)

P.T. Williams, Pyrolysis of waste tyres: A review. Waste Manag **33**, 1714–1728 (2013)

Chapter 3
Renewable and Sustainable Materials from Biological Approach

3.1 Microorganism as a Promising Sustainable Engineer

Microorganism is a tiny or microscopic organism and normally present in its single-cell form or in a colony. Microorganisms include bacteria, fungi, archaea, protists, algae and viruses. Basic nutrient is most important to the all microorganisms in the world for maintenance of metabolic functions and growth. Referring to the amount, type and concentration of nutrient, it totally depends on the metabolic pathway of the microorganism. Energy can be derived from carbohydrate, alcohols and amino acids by microorganism and most of the microorganisms have a capability to metabolise glucose as a simple sugar to get energy. Certain type of microorganisms can metabolise complex carbohydrates because those types of microorganisms produce enzyme to degrade the complex carbohydrate. Undeniably, microorganisms also are capable of producing materials with widely divergent chemical structures during their metabolisation process. Materials from microorganism can be divided into two types such as primary product and byproduct. Primary product is the main material been produced during the process of metabolisation. Meanwhile, byproduct is the secondary product derived during the process of metabolisation and reaction. Nowadays biotechnology field had been explored widely. Then, researchers from this field introduce technology of genetic engineering or also known as genetic modification, which is the direct manipulation of an organism's genes using biotechnology. This technology works with change the genetic makeup of cell with transfer genes within and across the limitations to produce improved or novel organism. Thus, the new DNA will be produced by isolating or copying the genetic material of interest using recombinant DNA methods or by artificially synthesising the DNA. Subsequently, the new DNA will be inserted into the host organism and an organism that is generated through genetic engineering is considered to be genetically modified (GM) and the resulting entity is a genetically modified organism (GMO). This technology is very useful in the industry to produce high yield of promising materials.

M. F. Yhaya et al., *Renewable and Sustainable Materials in Green Technology*,
SpringerBriefs in Applied Sciences and Technology,
https://doi.org/10.1007/978-3-319-75121-4_3

Microorganism can be considered as a sustainable producer because their process is natural process and does not disturb or destroy the environment. However, this sustainable process can be achieved if the men who are working for this process follows the procedure. For example, to produce methane from food waste via microorganism reaction, the most important is methane trap. If the methane trap is not working properly, this process is not only waste but it also will damage the environment because methane will contribute to greenhouse effect.

3.2 Overview of Material Production from Microorganism

Biological process technology is not a new technology as it has been applied a few decades ago. For instance, the production of methane or fuel for cooking through microorganism was recorded in India a few decades ago. Meanwhile, production of biodiesel through algae was a very common technique and has been practised in many countries. In this section, the processes to produce or recycle materials via biological approach are presented in detaile.

3.2.1 Bioconcrete and Biocement

Concrete is a composite material in which cement is one of the materials. Cement is a solid material made of clinker, gypsum and other additives, which is mainly used to form concrete and further involved in civil engineering constructions. Recently, biocement has slowly took over market of conventional cement and bioconcrete is with great development. Knowledge of construction microbial technology has been applied instead of using conventional method to produce biocement and bioconcrete. This technology involves microorganism as an 'engineer' to produce the sustainable biocement and bioconcrete. It is the invention of microorganism's involvement in carbonate precipitation. The emergence of biocement has solved the disadvantages that were coming from conventional method of cement production. As cement is one of the main materials to make concrete, the production of biocement is greatly related to the development of bioconcrete. The contents of the following have included the descriptions on the process of conventional method and its implication towards environment. The highlight of this report is on various microorganisms that are involved in the production of bioconcrete and biocement, and how it brings benefits to the environment.

Construction microbial biotechnology studies the microbially mediated construction processes and microbial production of construction materials. Specifically, the examples of construction processes applied with biotechnologies are bioaggregations, biocementation, bioclogging and biodesaturation of soil (Stabnikov et al. 2015). The examples of construction-related biomaterials produced from microorganisms are microbial cements, grouts, polysaccharides and bioplastics. Different microorganisms involved in the production of construction materials include bacteria, cyanobacteria, algae and also fungi (Ivanov et al. 2015).

Bioconcrete is a type of concrete created by mimicking the nature's way. The precipitation of minerals by living organisms occurs in nature environment as biomineralisation. For example, plants produce cystolith inclusion in a cellulose matrix, generally in the leaf of certain plants. Another example is the forming of bones, teeth and shells in animals (Gonsalves and Cuchí 2011). Beside plants and animals, microorganisms also have the ability to precipitate minerals such as calcites, carbonates, phosphates, oxides, sulphides, silicates, silver and gold. The biomineralisation process occurs at very slow rate naturally as in the formation of limestone, sandstone and others (Verma et al. 2015). Bioconcrete is modelled on microorganisms that can induce microbial carbonate precipitation (MCP). The precipitation is induced through the metabolic processes of the microorganisms in the presence of calcium precipitate calcium carbonate ($CaCO_3$). The organisms that can induce MCP are photosynthetic organisms, sulphate reducing bacteria and also organisms that involved in the nitrogen cycle (Gonsalves and Cuchí 2011).

Bacteria are the most applicable microorganisms for producing of bioconcrete. This is because bacteria are small in cell diameter, which ranges from 0.5 to 10 μm. They also have big physiological diversity such as pH 2–10 and temperature from −10 to 110 °C. There is also a big spectrum of biogeochemical reactions of bacteria that include oxidation–reduction of organics, oxygen, nitrate, ferric and sulphate. Bacteria also have highest growth and metabolic rate if compared to fungi, plants and animals (Ivanov et al. 2015). The production of bioconcrete materials generally includes two stages; the upstream process or fermentation, and also the downstream process. The upstream process is the preparation of medium, equipment and microbial inoculum. Downstream processes are processes after harvesting the product such as concentration of the product, drying, packaging, cleaning of the equipment and waste treatment (Ivanov et al. 2015). This material can be applied in repairing of concrete or to develop self-healing concrete. Self-healing materials will be discussed further in Chap. 4.

On the other hand, there are several types of biocementation which depend on the type of metabolic processes carried out in bacteria and a few will be discussed in this chapter. One of them is calcium and urea-dependent biocementation. It is based on microbially induced calcium carbonate precipitation (MICCP), which is the formation of calcium carbonate minerals (Ivanov et al. 2015). This is the most commonly used approach for production of calcium carbonate and bacteria with urease activity can be easily found (Gonsalves and Cuchí 2011). The processes involved are shown in the Eqs. 3.1, 3.2, 3.3, 3.4 and 3.5 (Ivanov et al. 2015).

$$(NH_2)_2\,CO + 2H_2O \xrightarrow{\text{Urease}} CO_2 \uparrow +2NH_4OH, \tag{3.1}$$

$$CO_2 + H_2O \leftrightarrow H_2CO_3 \xrightarrow{\text{Carbonicanhydrase}} H^+ + HCO_3^- \leftrightarrow 2H^+ + CO_3^{2-}, \tag{3.2}$$

$$CaCl_2 + H_2CO_3 \rightarrow CaCO_3 \downarrow +2HCl, \tag{3.3}$$

$$2HCl + 2NH_4OH \rightarrow 2NH_4Cl + 2H_2O, \tag{3.4}$$

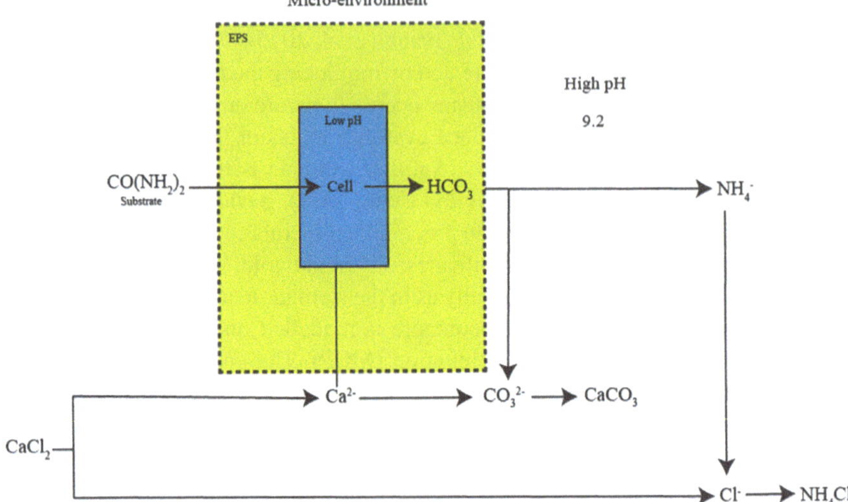

Fig. 3.1 Schematic representation summarising the different phases during $CaCO_3$ precipitation process via ureolytic bacteria

Total

$$(NH_2)_2\,CO + 2H_2O + CaCl_2 \xrightarrow{\text{Urease and carbonicanhydrase}} CaCO_3 \downarrow + 2NH_4Cl. \tag{3.5}$$

From these equations, we can see that there are several processes involved in the calcium and urea-dependent biocementation. The hydrolysis of urea generates carbonate ions at 1:1 molar ratio. Another process is the surface adsorption of calcium ions. The calcium ions are supplied in the form of calcium chloride, in which the ions are attracted to the negatively charged bacterial cell wall (Gonsalves and Cuchí 2011). The calcium ion will then deposit on the cell surface. The cementation process is shown in the Eqs. 3.6, 3.7 and 3.8 (Gonsalves and Cuchí 2011).

$$Ca^{2+} + Cell \rightarrow Cell.\,Ca^{2+} \tag{3.6}$$

$$Cl^- + HCO_3^- + NH_3 \leftrightarrow NH_4Cl + CO_3^{2-} \tag{3.7}$$

$$Cell.\,Ca^{2+} + CO_3^{2-} \rightarrow Cell.\,CaCO_3 \downarrow \tag{3.8}$$

Another process is the nucleation and crystal growth. When the solution is saturated with the solute, the solubility of calcium carbonate is low and it will precipitate. The formation of new crystals begins at the point of critical saturation. The nucleation sites for formation of crystals can be on the bacterial cells, and once the nucleus is stable the crystal grows spontaneously (Gonsalves and Cuchí 2011). Figure 3.1 shows the whole process of $CaCO_3$ by ureolytic bacteria.

The second type is based on the production of carbonates by heterotrophic bacteria during aerobic or anoxic oxidation of organics. During aerobic oxidation of organics, carbonate is produced and pH increases, this may result in the precipitation of calcium carbonate, for example as shown in the Eq. 3.9 (Ivanov et al. 2015).

$$(CH_3COO)_2\,Ca + 4O_2 \rightarrow CaCO_3 \downarrow + 3CO_2 \uparrow + 3H_2O. \quad (3.9)$$

The precipitation of calcium carbonate can also be obtained from the anoxic oxidation of organics by heterotrophic bacteria, for example the bioreduction of nitrate as shown in the Eq. 3.10 (Ivanov et al. 2015).

$$(CH_3COO)_2\,Ca + \frac{8}{5}NO_3^- \rightarrow CaCO_3 \downarrow + \frac{4}{5}N_2 \uparrow + 3CO_2 \uparrow + 3H_2O + \frac{8}{5}OH^- \quad (3.10)$$

The third type of reaction is calcium phosphate biocementation. The calcium phosphate precipitation from calcium phytate solution is using phytase activity of microorganisms to produce a mixture of crystal forms such as monetite, whitlockite and hydrocyapatite. Triple superphosphate can also be used for calcium phosphate precipitation using acid-tolerant urease producing microorganisms. The reaction described is shown in the Eqs. 3.11 and 3.12 (Ivanov et al. 2015).
Monetatite precipitation:

$$Ca(H_2PO_4)_2 + CO\,(NH_2)_2 + H_2O + \text{acid urease} \rightarrow CaHPO_4 \downarrow + CO_2 + (NH_4)_2\,HPO_4^- \quad (3.11)$$

Hydroxyapatatite precipitation:

$$5Ca(H_2PO_4)_2 + 8CO\,(NH_2)_2 + 8H_2O + \text{acid urease} \rightarrow Ca_5\,(PO_4)_3\,(OH) \downarrow + 2NH_4HCO_3 + 6CO_2 + 7\,(NH_4)_2\,HPO_4 \quad (3.12)$$

3.2.2 *Polyhydroxyalkanoates (PHA)*

Polyhydroxyalkanoates (PHA) are polyesters made up of numerous hydroxyalkanoates, which are produced by various gram-positive and gram-negative bacteria under unbalanced growth conditions. Microbes produce PHA and store it intracellularly as carbon and energy reserve under nutrients limitation such as depletion of nitrogen, sulphate, phosphate, magnesium and oxygen but in excess of carbon. Carbon source is a very important factor in mass production of PHA as excess carbon will be stored in the form of PHA granules and is used during starvation period (Chen and Wu 2015; Reddy et al. 2003).

Besides, there are various microbes able to produce PHA such as *Pseudomonas*, *Bacillus*, *Ralstonia*, *Aremonas*, *Rhodobacter* and *halobactericeae*. PHA particularly include poly 3-hydroxybutyrate (PHB), copolymers of 3-hydroxybutyrate and 3-hydroxyvalerate (PHBV), poly 4-hydroxybutyrate (P4HB), copolymers of 3-hydroxybutyrate and 3-hydroxyhexanoate (PHBHHx) and poly 3-hydroxyoctanoate (PHO) which are widely used in numerous applications. PHA exists as discrete inclusions that are typically 0.2–0.5 μm in diameter localised in the cell cytoplasm and may be visualised quite clearly with a phase contrast light microscope due to their high refractivity (Philip et al. 2007; Sudesh et al. 2017).

In addition, there are more than 100 different monomer units that have been identified as constituents of the storage PHA. In other words, different types of biodegradable polymers could be produced over an extensive range of properties. The molecular mass of PHA varies with the PHA producer which is between the range of 50–1000 kDa. PHA can be classified into short-chain-length (SCL) and medium chain length (MCL), which depend on the number of carbon atoms of the monomer. The SCL polymer consists of 3–5 carbon atoms, whereas MCL polymer possesses 6–14 carbon atoms of the monomer (Philip et al. 2007).

A range of PHAs with desirable properties can be obtained based on the copolymer composition of the PHA where these polymers can be hard and crystalline or elastic and rubbery. First, poly(3-hydroxybutyrate) [P(3HB)] is highly crystalline and melts at 180 °C with the properties of brittleness and stiffness. Therefore, the introduction of different HA monomers such as 3-hydroxyvalerate (3HV) or 3-hydroxyhexanoate (3HHx) into the chain greatly improves the material properties of P(3HB). P(3HB) is a biocompatible polymer, is optically pure and possesses piezoelectricity properties.

Besides, poly(3-hydroxybutyrate-co-3-hydroxyvalerate), P(3HB-3HV) is another example of PHA which has a lower melting temperature and lower crystallinity than P(3HB). Determination of copolymer's properties is highly related to the mole percentage of 3HV in the polymer. Moreover, P(3HB-3HV) consists of more than 20 mol% of 3HV units can be used to make films and fibres with different elasticity by controlling the processing conditions.

Poly(4-hydroxybutyrate), P(4HB) is a strong and malleable thermoplastic material with a tensile strength similar to commercial polyethylene where it has an extremely elastic properties. When combined with other hydroxyacids, the material properties of P(4HB) can be varied (Chen and Wu 2015; Philip et al. 2007).

PHA has been widely used in many applications due to its novel features. At first, PHA is used in packaging films mainly in bags, containers and paper coatings. In addition, there are similar applications as conventional commodity plastics include the disposable items, such as razors, utensils, diapers, feminine hygiene products, cosmetic containers shampoo bottles and cups. Generally, PHA is considered as a source for the synthesis of chiral compounds (enantiomerically pure chemicals) and act as raw materials for the production of paints. PHA can be easily depolymerised to a rich source of optically active, pure, bi-functional hydroxy acids. PHB, for instance is readily hydrolysed to R-3-hydroxybutyric acid and used in the synthesis of Merck's anti-glaucoma drug 'Truspot'. In tandem with R-1,3-butanediol, it is also used in the synthesis of β-lactams.

Furthermore, PHA is also useful as stereo-regular compounds, which can serve as chiral precursors for the chemical synthesis of optically active compounds. These compounds can be used as biodegradable carriers for long-term dosage of drugs, medicines, hormones, insecticides and herbicides. They are also used as osteosynthetic materials in the stimulation of bone growth owing to their piezoelectric properties, in bone plates, surgical sutures and blood vessel replacements. However, the medical and pharmaceutical applications are limited due to the slow biodegradation and high hydraulic stability in sterile tissues.

A lot of microorganisms are able to produce PHA with various characteristics and properties. Genetically engineered microbes also used to produce PHA, for example, recombinant *E.coli* is able to utilise molasses to produce PHA in batch culture as reported (Saranya and Shenbagarathai 2011). *Burkholderia cepacia* belongs to gram-negative bacteria, it was first isolated by William Burkholder back in 1950. At first, it was identified as *Pseudomonas* due to the similar phenotypes, only in 1992 it was reclassified as a new genus and was renamed as *B.cepacia* (Mahenthiralingam et al. 2011). *B.cepacia* is revealed to have high ability to produce high PHA concentration under nutritional limitation. *B.cepacia* is known for its ability to produce different short-chain-length (SCL) PHAs such as PHB, PHV and PHB-co-PHV. It can utilise various carbon sources such as xylose, galactose, glucose, levulinic acid and glycerol to grow and encourage PHA production (Zhu et al. 2013).

As mentioned above, PHAs are produced by the bacteria under the condition where there are abundant carbon sources and limited nutritional sources. *B.cepacia* is reported to be able to utilise different carbon sources to produce PHAs including xylose, galactose, glucose, glycerol and levulinic acid (Keenan et al. 2004). In this case, glycerol is discussed here. Glycerol is generally transformed into P(3HB) under such unfavourable conditions.

First of all, glycerol is converted into glyceraldehyde-3-phosphate (GAP) by three enzymes namely: glycerol kinase (GlpK), glycerol-3-phosphate dehydrogenase (GlpD) and triosephosphate isomerase (TPI). GAP is then metabolised to pyruvate, which is later converted into acetyle-CoA in pyruvate decarboxylation. Microbial biosynthesis of P(3HB) starts with condensation of two molecules of acetyl-CoA by b-ketothiolase (PhaA). The condensed acetyl-CoA is subsequently reduced to (R)-3-hydroxybutyryl-CoA by acetoacetyl-CoA reductase (PhaB). Finally, PHA synthase (PhaC) polymerized 3-hydroxybutyryl-CoA moieties to P(3HB). The process is shows in Fig. 3.2 (Zhu et al. 2013).

3.2.3 1-3 Propanediol (PDO)

1-3 propanediol known as PDO is a monomer and it is applied in many industries. It also has potential l utility in the production of polyester fibres. It is used widely in the industries such as cosmetics, lubricants and medicines (Drożdżyńska et al. 2011; Menzel et al. 1997). It is also known as a trimethylene glycol, 1.3-dihydroxypropane, propane-1,3-diol. The compound has a chemical formula of $C_3H_8O_2$ with molecular mass 76.09 g mol^{-1}. Then the boiling point is 210–212 °C, meanwhile the melting point is 28 °C (Igari et al. 2000).

The production of PDO can be divided into two methods of processes. First, through chemical process and other is fermentation or biological process. Chemical process is a very common practice in this industry. This process will use ethylene oxide phosphine, carbon monoxide, hydrogen and acid. However, this process

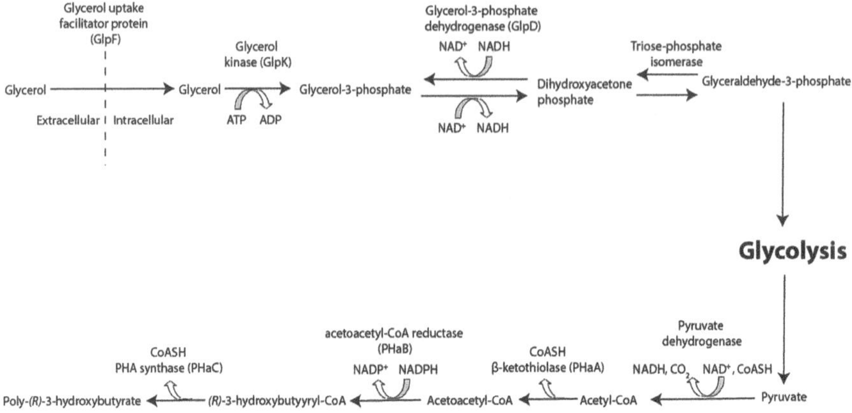

Fig. 3.2 Synthesis pathway of Poly-(R)-3-hydroxybutyrate from glycerol

requires high pressure, high temperature and catalyst. The material such as ethylene oxide comes from fossil fuel. This process is quite expensive and it will lead to high cost of production and finally impacts the price of PDO in the market. Besides this process also will generate hazardous waste and will not be friendly to environment and require cost for treatment.

Meanwhile the fermentation process to produce PDO in industrial scale by microorganism is not a new technology, because it has been discovered and applied many years ago. Currently, in the industry scale they use glucose from corn as the main media for PDO production. After the process of harvesting corn from the plantation, this crop will undergo milling and separating operations. The advantages of milling are to reduce the size, increase the surface area, and so following processing steps are more effective. Kernels from corn will be cooked (steam explosion) for around 24 h at 125 °C. This process is to ensure kernels are swollen and soften. After that, specific enzyme will be added in the swollen and soften kernels. This enzyme will convert starch from the soft kernels to become glucose and this process known as hydrolysation. After that, hydrolysed media is ready go to next process, that is fermentation (Kurian 2005).

The process of fermentation can be divided into two phases. Both phases use recombinant microorganism, which is the microorganism that goes through the phases of genetic modification (Celińska 2010; Nevoigt and Stahl 1997; Saxena et al. 2009). The first phase is to convert glucose to glycerol and the microorganism used is recombinant yeast. In the second phase, glycerol will be converted to PDO by using recombinant *E.coli*.

During the process of fermentation specific value of temperature, aeration, time and agitation will be applied due to high yield of PDO. This value is depending on the type of microorganism has been selected. However, a few researchers have been reported that recombinant *E.coli* has high yield and almost 99.7% of purity after the process of refinery (Kurian 2005). Then after the process of fermentation is complete,

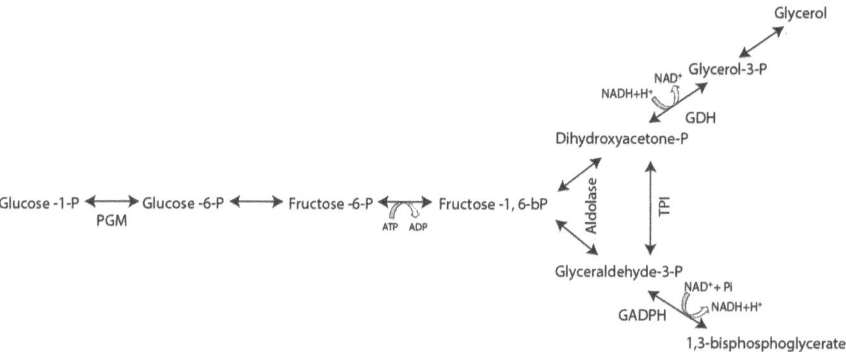

Fig. 3.3 Conversion of glycerol from glucose-1-phosphate

the fermented media will go to the process refining which is a process to harvest or collect PDO. The process of downstream that can be applied to refinery PDO are filtration, sorption, evaporation and distillation. The biological process to produce PDO can be considered environmental friendly, in which the waste generation is not considered as hazardous waste. This process could be considered better than chemical process because it uses low temperature and pressure. However the problem in this process is the main media for fermentation is corn, since corn is the staple food for many parts of the world (Kurian 2005). In a report of DuPont and Genecor International Inc., a metabolically engineered *E. coli* could produce up to 135 g/l of 1,3-PD with a yield of 0.6 mol 1,3-PD/mol glucose (Nakamura and Whited 2003; Saxena et al. 2009).

The innovation of bio-PDO has reduced the dependence on fossil fuels and it is economically more favourable than fossil-derived propanediol. Many researchers noticed that bio-PDO consumes 40% less energy in terms of line production and reduces more than 40% greenhouse gas emissions. Figure 3.3 shows the conversion of glycerol to glucose and Fig. 3.4 shows the conversion PDO from glycerol by recombinant yeast and *E.coli*.

Many industries are interested in microbial-based 1,3-PD production, as it appears to be competitive as compared to traditional technologies. 1,3-PD is a very useful bulk chemical, with its wide array of applications and benefits to human beings. For instances, it is used in the manufacture of polymers, cosmetics, food, lubricants, and medicines. 1,3-PD can be from patent microorganism such as recombinant *E.coli.* The biotechnological method has advantages compared to the traditional chemical production. This is due to chemical and traditional synthesis of 1,3-PD which requires high temperature, pressure, catalyst and generates waste stream that can cause environmental pollution.

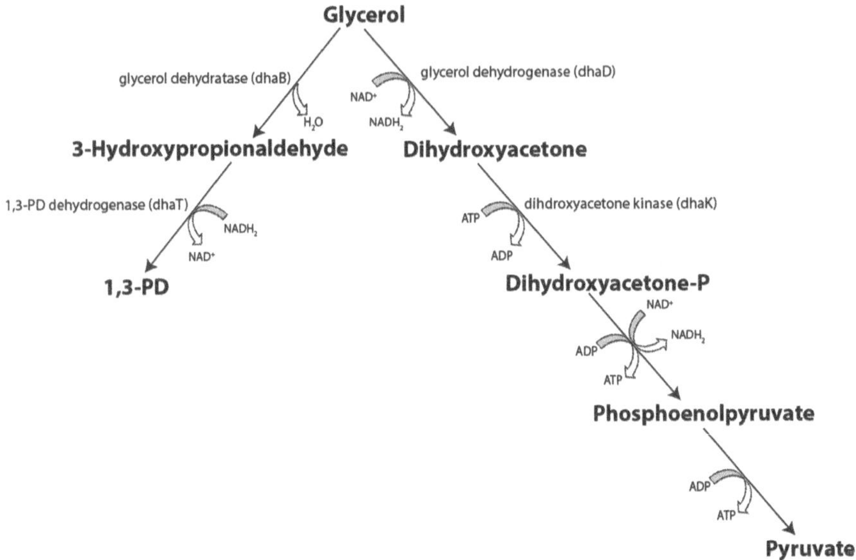

Fig. 3.4 Conversion of PDO from glycerol

3.3 Upstream Processing

Upstream and downstream processing are involved in the process to produce materials from microorganism. Upstream is a process which starts from early cell isolation, cultivation and operation of bioreactors. Upstream is very important for biological process to optimise the production or to increase the yield and simultaneously reduce the cost of operation. In industrial scale, operation of bioreactor becomes their main concern because this leads to cost-effectiveness and yield production.

Normally, in the industrial scale, bioreactors are used to cultivate microorganism for growth and guide the development of produce materials because it easy to work in the industrial scale, and is controllable and a close system. The whole cells of microorganism transform raw materials into biochemical products and or less undesirable byproducts into the bioreactors' vessels or tanks. Bioreactors also can be classified as in vitro culture systems which have been planned to alter the following basic physiological phenomena, cell survival, organisation, mechanical properties, production and function. Bioreactors ensure cell survival through adequate delivery of essential nutrients produce the materials such as their capability.

3.3.1 Modes of Bioreactor

The design and operation of bioreactors (modes of bioreactor) are keys to maximising productivity and reliability. The requirements of a good reactor design are many fold but the key aspects are to:

- Provide an ideal environment for growth
- Provide an environment for safe and robust operations
- Exploit and enhance product formation kinetics and yield
- Make a simple cost-effective system

The reactors for the growth of clostridia can be relatively simple and originally were developed from brewing type systems using batch reactors. This is quite appropriate as many of the important products produced are not coupled directly to growth and are either non-(or slow) growing end product inhibited conditions or are non-growth-related secondary products. As such batch systems provide an effective production system. Continuous culture systems in many cases are unsuitable because the maintenance of the cells in such systems is related directly to the yield and growth of the organisms, hence are not suitable for end product inhibition or non-growth-related product formation.

There is a large body of work investigating reactor designs for *C. acetobutylicum* and related organisms that have the capability to produce solvents, ethanol acetone and butanol undergrowth inhibitory or non-growing conditions (Zverlov et al. 2016: Cheng et al. 2012). However, the downside of a batch process is that it is a relatively low productivity system as compared with continuous systems or systems containing high biomass concentration such as packed-bed reactors or reactors with cell recycle.

(a) *Batch Operation*

In a batch reactor, the reactor is filled with organic matter and microorganisms (a batch) and the process of decomposition is allowed to proceed for a predetermined time or until gas production decreases to a predetermined (low) rate. Normally, in this type of operation, 10–20% of the material is left as a seed when the reactor is reloaded and the operation repeats (Chongrak 1996). Many researchers have reported the application of this operation to culture *C. Butyricum* (Abbad-Andaloussi et al. 1998; Wang and Jin 2009; Vandak et al. 1997).

In batch cultures, the growth of microorganism will pass through a number of phases and typically starts with the lag phase in which the cells adapt to their new environment. The specific growth rate (μ) then slowly accelerates until it reaches the log or exponential phase. In this phase growth is only limited by the capacity of biomass to grow [the maximum specific growth rate (μ_{max}) and the cell concentration (x)]. When the microorganisms have been growing in the exponential phase for some time, nutrient depletion and possible end product inhibition will occur causing the growth rate to slow down and finally stop as the culture enters the stationary phase. Here, the growth rate (μ) for the microorganism is equal to the death rate (k_d) of the

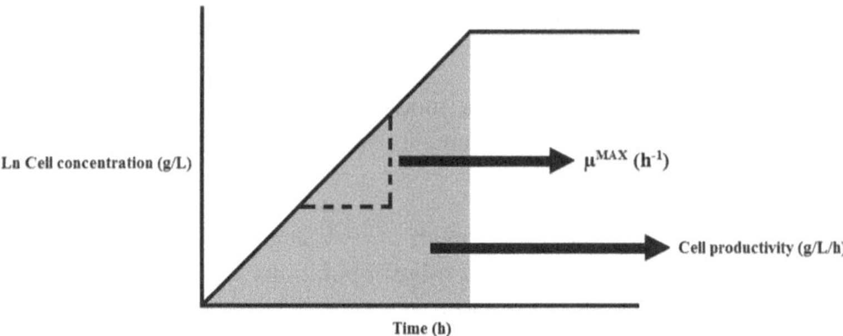

Fig. 3.5 Cell concentration versus time in the batch system. Cell productivity will be dependent on the period in which the culture is harvested

microorganism so $dx/dt = 0$. Sometime after the stationary phase, the microorganism enters the death phase where there is a net loss of active biomass.

Figure 3.5 shows the graph for the concentration of cell versus time. The shaded area is the cell productivity (g/L/h) and the slope of the graph gives $\mu(h^{-1})$.

(b) *Continuous Operation*

The main concept of the continuous system is that it is an open system where sterile substrate feed is continuously added into the bioreactor. The feeding rate of the substrate is equal to the amount of converted nutrient solution with microorganisms that is simultaneously removed from the system. In a continuous process under steady-state conditions, cell loss as a result of outflow must be balanced by outgrowth of the organism. Continuously operating reactor systems are very well suited to the treatment of liquid and organic waste (with low solid concentration). These processes typically start as a batch process and once the microbial population establishes itself a feed pump is started and the production stabilises at steady state where the growth rate of biomass is equal to that leaving the reactor (Chongrak 1996). The disadvantage of this system is the possibility of contamination due to the fact that it is an open system. The sterility of the feed is of prime importance. Figure 3.6 shows cell concentration versus time for a continuous system. The shaded area is cell productivity and dilution rate D (h^{-1}) is equal to growth rate, $\mu(h^{-1})$. In these systems, productivity is much higher than observed in batch culture as the system is operating at steady state.

3.3.2 *Type of Bioreactor*

(a) *Conventional Bioreactor*

Stirred tank bioreactor is equipped with an agitator to mix the reactants, and integrated with heating and cooling system. This bioreactor is also equipped with the help to mixing and sparger for aeration (Singh et al. 2014). Stirred tank type of bioreactor, which is also known as stirred tank reactor (STR), is the most common type

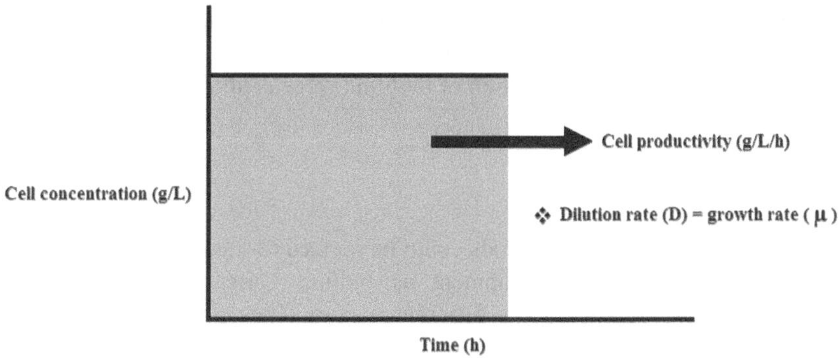

Fig. 3.6 Cell concentration versus time in the continuous system

of fermenter used today. This unit offers low capital and operating cost (Williams 2002). Williams (2002) also explained the process involving continuous stirred tank reactor (CSTR) in which substrate is continuously fed in bioreactor while the product is drawn from the fermenter to balance the reaction.

Air-lift bioreactor uses air as a principal means to achieve various transport requisites. The most important things are that air-lift bioreactor offers simplicity and low capital investment when compared to the mechanically stirred bioreactor (Chen 1990). The author also added that air-lift fermenter has higher efficiency compared to mechanical stirred at the same power input. This bioreactor is suitable for large-scale production. Air-lift bioreactor is also known as tower bioreactor. Despite the advantages, there is also a few disadvantage of air-lift fermenter, which is the inefficiency of gas and liquid separation when foaming occurs. Foaming is a big issue when dealing with fermentation as it would lower the reaction and result in loss of media (Williams 2002).

Packed-bed bioreactor consists of a bed of packing made of polymer, ceramic, glass or natural material which comes in a variety of shapes and sizes (Singh et al. 2014). The core principle behind the design of this bioreactor is that cells are immobilised within the matrix that forms the bed. By doing so, the surface area provided for the cell attachment is high, resulting in a concentration of cell as high as 5×10^8 cells m^{-1} of the matrix (Warnock et al. 2005).

Singh et al. (2014) reported that the fluid containing nutrient flow through the solid bed and the nutrient flow rate can be controlled to increase or decrease contact time. The main problem with the packed-bed reactor is heterogeneity of the bioreactor caused by concentration nutrient and waste product (Warnock et al. 2005).

Fluidised bed bioreactor is similar to the packed-bed bioreactor but with smaller particle size. Thus, fluidised bed bioreactor can eliminate the problem encountered in packed-bed bioreactor such as clogging, bed compaction and high liquid pressure drop (Singh et al. 2014). Since fluidised bed bioreactor is a continuous reactor, thus it shares a common problem which hasa high risk of contamination, genetic instability and back mutation (Andrews 1988). The growth support media retained in suspension by drag forces exerted by the flowing medium resulting in uniform contact and maximised the formation of biofilm (Perez et al. 2001)

(b) *Modification bioreactor*

(i) *Immobilised cell reactor*

In this type of reactor, three methods could be applied to immobilise the cells such as covalent bond formation, entrapment and biofilm (Qureshi et al. 2005). All of these methods have been applied in different reactors. The types of immobilised cell reactor include;

Fluidised bed reactors with immobilised cell—This reactor operates with an upward flow of liquid. It is simpler than the packed-bed reactor. Fluidised bed reactors are often used in waste treatment with sand or another suitable carrier material which supports the mixed microbial populations inside the fluidised bed. Other applications for this type of reactor include the commercial production of vinegar. In this process, a flocculating organism is applied.

Packed back reactors with immobilised cell—This kind of reactor is used commercially with immobilised cells and enzymes for the production of aspirate and fumarate, conversion of penicillin to 6 amino penicillanic acids and resolution of amino acid isomers. Normally the medium (continuous liquid flow) will feed through a packed-bed tube either at the top or bottom of the reactor tube. Inside, the packed bed is equipped with catalyst particles (enzymes or cells). The advantage of this kind of reactor is that damage to the enzymes or cells or fermentation is lower than in a comparable stirred reactor.

Trickle bed reactor with immobilised cell—The trickle bed reactor is similar to a fluidised bed reactor. The difference being that the trickle bed reactor is fitted with a spray distributor. The liquid is recycled from the bottom of the reactor to the top of reactor by spraying it back through the spray distributor into the top of the reactor. This reactor also has the capability to produce hydrogen by *C. acetobutylicum* with high of yield (Zhang et al. 2006).

Air-lift reactor with immobilised cell—The concept of the air-lift reactor is similar to that of the bubble column reactor. However, it is equipped with a draft tube. Air-lift reactors are often chosen for the immobilised catalyst, cultural animal and plant cell digestion because shear levels are lower than stirred vessels. The resulting fraction of distributed cultivations is very low.

(ii) *Cells recycle system*

If cells are retained in the reactor system then enhanced reactor performance over classic continuous cultures is possible. Fermentation processes whereby cells are retained in the reactor by sedimentation to give very high cell concentrations result in higher volumetric catalytic power leading to high productivity. This allows higher flow rate through the system well above that for cells in CSTR. This type of enhancement exploits common behaviour in natural environments to form rapidly settling flocks found in fresh and seawater environments. These systems, therefore, can naturally be selected for flock communities as they are retained over equivalent free-living organisms. There are many variations developed and used especially in waste treatment, from the classic activated sludge plant to upward-flow anaerobic sludge blanket (USAB) reactors. This type of reactor has not only been applied widely in the wastewater treatment, but also has been applied to the investigation of acetic acid production by *C. thermoaceticum* (Schwartz and Keller 1982).

MBR (Membrane bioreactor)—The main concept for this reactor is to retain cells in the reactor by filtration. There are two types of basic design in these systems depending on the use of the membrane. The first type used membranes which filter water by suction through the membrane surface removing clarified water while retaining the particulates. This type of system is commonly used in waste treatment as a replacement for activated sludge processes or flocculation processes where filtration replaces sedimentation to retain biomass within the reactor. The second type of MBR filters pressurise water to force water out of the system. These systems are easier to engineer especially when sterile environments are required in systems run under positive pressure. They are more often used in reactors with pure cultures. MBR systems have many advantages over continuous culture or reactors with cell recycle relying on sedimentation, in that cell retention is controlled by a physical separation and so can be generically applied to all types of cells to produce high cell concentrations in the reactor and so allow high productivities. In addition, permeates from MBRs are cell free, thus allowing their further separation and purification. Lactic acid productivity ($P_{x/t}$) and final biomass concentration by lactic acid bacteria in the MBR were over 20 times greater that which is better than those for stirrer reactor (STR) (Jung and Lovitt 2010).

3.4 Downstream Processing

Downstream is a process to distress with the extraction of desired products from the upstream biomass. Downstream could be classified as the process of removal insoluble, isolation, purification and polishing of product from fermented broth.

In many cases, production of materials from biological process may be considered to combine two or more stages, for example as mentioned above the process of downstream to produce PDO is started from filtration, sorption, evaporation and finally distillation.

3.4.1 Removal of Insoluble

Removal insoluble means the process separation of cells, cell debris or other particulate matter. In the certain case if a product or material needs to be harvested is intercellular. Intracellular means that material been produced inside the cell, for example is PHA. Then the process of cell disruption is needed. This process also could be considered as removal of insoluble. For the extracellular product or materials, probably this method is not required and it just needs to process separation product from cell. For example, the production of $CaCO_3$ in the industry bioconcrete and biocement. Normally to remove of insoluble, many techniques could be considered such as filtration, centrifugation, sedimentation, flocculation, gravity settling and cell disruption.

3.4.2 Product Isolation

Next step is product isolation. Product isolation means the techniques to remove unwanted or impurity and the head of the impurity in the biological process is water. The process isolation product included reducing volume for concentrated the product. The example of a technique for isolation product is liquid–liquid extraction, adsorption, filtration (ultrafiltration), and precipitation.

3.4.3 Product Purification

Some of materials produced from biological process not required to process of purification. However certain materials need this process. This process is expensive to carry out. This process also needed sophisticated and sensitive requirement. The example techniques for product purification are crystallisation, chromatography and fractional precipitation.

3.4.4 Polishing

This is the last stage of downstream processing before the material going to process packaging. The aim of this stage is to ensure that material or product stable and easily to handle during the process packaging and transportation. The example technique in this stage is spray drying, sterilisation and desiccation.

References

S. Abbad-Andaloussi, J. Amine, P. Gerard, H. Petitdemange, Effect of glucose on glycerol metabolism by *Clostridium butyricum* DSM 5431. J. Appl. Microbiol. **84**(4), 515–522 (1998)

G. Andrews, Fluidized-bed Bioreactors. Biotechnol. Genet. Eng. Rev. **6**, 151–178 (1988)

E. Celińska, Debottlenecking the 1, 3-propanediol pathway by metabolic engineering. Biotechnol. Adv. **28**(4), 519–530 (2010)

N.Y. Chen, The design of airlift fermenters for use in biotechnology. Biotechnol. Genet. Eng. Rev. **8** (1990)

G.Q. Chen, Q. Wu, The application of polyhydroxyalkanoates as tissue engineering materials. Biomaterials **26**(33), 6565–6578 (2015)

C.L. Cheng, P.Y. Che, B.Y. Chen, W.J. Lee, C.Y. Lin, J.S. Chang, Biobutanol production from agricultural waste by an acclimated mixed bacterial microflora. Appl. Energy **100**, 3–9 (2012)

P. Chongrak, *Organic waste recycling*. John Willey and Son Inc. USA (1996)

A. Drożdżyńska, K. Leja, K. Czaczyk, Biotechnological production of 1,3-propanediol from crude glycerol. BioTechnologia. **92**(1), 92–100 (2011)

G.M. Gonsalves, A.B. Cuchí, *Bioconcrete-a sustainable Substitute for Concrete?* (Universitat Politecnica, De Catalunya, Barcelona, 2011), p. 3

V. Ivanov, J. Chu, V. Stabnikov, *Chapter 2 basics of construction microbial biotechnology* (Springer International Publishing Switzerland, 2015), pp. 1–37

I. Jung, R.W. Lovitt, A comparative study of the growth of lactic acid bacteria in a pilot scale membrane bioreactor. J. Chem. Technol. Biotechnol. **85**(9), 1250–1259 (2010)

T.M. Keenan, S.W. Tanenbaum, A.J. Stipanovic, J.P. Nakas, Production and characterization of Poly-a-hydroxyalkanoate copolymers from *Burkholderia cepacia* utilizing xylose and levulinic acid. Biotechnol. Prog. **20**(6), 1697–1704 (2004)

J.V. Kurian, A new polymer platform for the future Sorona® from corn derived 1, 3 propanediol. J. Polym. Environ. **13**(2), 159–167 (2005)

E. Mahenthiralingam, A. Baldwin, C.G. Dowson, Burkholderia cepacia complex bacteria: opportunistic pathogens with important natural biology. J. Appl. Microbiol. **104**(6), 1539–1551 (2011). https://doi.org/10.1111/j.1365-2672.2007.03706.x

K. Menzel, A.P. Zeng, W.D. Deckwer, High concentration and productivity of 1,3-propanediol from continuous fermentation of glycerol by Klebsiella pneumonia Enzyme and Microbial Technology. **20**(2), 82–86 (1997)

C.E. Nakamura, G.M. Whited, Metabolic engineering for the microbial production of 1, 3-propanediol. Curr. Opin. Biotechnol. **14**(5), 454–459 (2003)

E. Nevoigt, U. Stahl, Osmoregulation and glycerol metabolism in the yeast Saccharomyces cerevisiae. FEMS Microbiol. Rev. **21**(3), 231–241 (1997)

M. Perez, L.I. Romero, D. Sales, Organic matter degradation kinetics in an anaerobic thermophilic fluidised bed bioreactor. Anaerobe **7**(1), 25–35 (2001). https://doi.org/10.1006/anae.2000.0362

S. Philip, T. Keshavarz, I. Roy, Polyhydroxyalkanoates: biodegradable polymers with a range of applications. J. Chem. Technol. Biotechnol. **82**(3), 233–247 (2007)

N. Qureshi, B.A. Annous, T.C. Ezeji, P. Karcher, I.S. Maddox, Biofilm reactors for industrial bioconversion processes: employing potential of enhanced reaction rates. Microb. Cell Fact. **4**, 24 (2005)

C. Reddy, R. Ghai, V.C. Kalia, Polyhydroxyalkanoates: an overview. Biores. Technol. **87**(2), 137–146 (2003)

S. Igari, S. Mori, Y. Takikawa, Effects of molecular structure of aliphatic diols and polyalkylene glycol as lubricants on the wear of aluminium. Wear. **244**, 180–184 (2000)

V. Saranya, R. Shenbagarathai, Production and characterization of PHA from recombinant *E. coli* harbouring phaC1 gene of indigenous *Pseudomonas* sp. LDC-5 using molasses. Braz. J. Microbiol. **42**(3), 1109–1118 (2011). https://doi.org/10.1590/S1517-83822011000300032

R. Saxena, P. Anand, S. Saran, J. Isar, Microbial production of 1, 3-propanediol: recent developments and emerging opportunities. Biotechnol. Adv. **27**(6), 895–913 (2009)

R.D. Schwartz, F.A. Keller Jr., Acetic acid production by *Clostridium thermoaceticum* in pH-controlled batch fermentations at acidic pH. Appl. Environ. Microbiol. **43**(6), 1385–1392 (1982)
J. Singh, N. Kaushik, S. Biswas, Bioreactors—technology & design analysis. The Scitech J. **01**(06) (2014)
V. Stabnikov, V. Ivanov, J. Chu, Construction Biotechnology: a new area of biotechnological research and applications. World J. Microbiol. Biotechnol. **31**, 1303–1314 (2015). https://doi.org/10.1007/s11274-015-1881-7
K. Sudesh, H. Abe, Y. Doi, Synthesis, structure and properties of polyhydroxyalkanoates: biological polyesters. Prog. Polym. Sci. **25**(10), 1503–1555 (2017)
D. Vandak, J. Zigova, E. Sturdik, S. Schlosser, Evaluation of solvent and pH for extractive fermentation of butyric acid. Proc. Biochem. **32**, 245–251 (1997)
R.K. Verma, L. Chaurasia, V. Bisht, M. Thakur, BioMineralization and bacterial carbonate precipitation in mortar and concrete. Biosci. Bioeng. **1**(1), 5–11 (2015)
X. Wang, B. Jin, Process optimization of biological hydrogen production from molasses by a newly isolated *Clostridium butyricum*. J. Biosci. Bioeng. **107**(2), 138–144 (2009)
J.N. Warnock, K. Bratch, M. Al-Rubeai, Packed bed bioreactors, in *Bioreactors for Tissue Engeneering*, ed. by M. Al-Rubeai, J. Chaudhuri (Springer, 2005), pp. 87–113
J.A. Williams, Keys to bioreactor selections. Chem. Eng. Prog. (2002)
H. Zhang, M.A. Bruns, B.E. Logan, Biological hydrogen production by *Clostridium acetobutylicum* in an unsaturated flow reactor. Water Res. **40**, 728–734 (2006)
C. Zhu, S. Chiu, J.P. Nakas, C.T. Nomura, Bioplastics from waste glycerol derived from biodiesel industry. J. Appl. Polym. Sci. **130**(1), 1–13 (2013). https://doi.org/10.1002/app.39157
V.V. Zverlov, O. Berezina, G.A. Velikodvorskaya, W.H. Schwarz, Bacterial acetone and butanol production by industrial fermentation in the Soviet Union: use of hydrolyzed agricultural waste for biorefinery. Appl. Microbiol. Biotechnol. **71**, 587–597 (2016)

Chapter 4
Renewable and Sustainable Materials for Various Green Technology Applications

4.1 Overview Materials Production

Materials, including those renewable and sustainable for green technology applications, can be produced through either single unit or integrated processes. Single unit refers to one or more grouped operations from similar processes. On the other hand, integrated process or production is a term referring to a holistic approach which emphasises the unity of the process or production and interaction between different unit operations. It combines more than one specific unit operation or process into a single piece of equipment or into a group of workstations that are operated under unified control (NRC 1995). This approach can combine multiple processes either chemical or biological methods as previously described in Chaps. 2 and 3, either in similar process or in different process families.

With the increasing attention of multidisciplinary research and development gearing towards innovative solutions, the integrated process or production becomes necessary in both research and education for the development of new knowledge and novel processes, especially in areas where multidisciplinary research and development is a must. There are a lot of factors that have pushed forward this integrated approach including the demand for reducing costs of and process, shorter timeframe of processing and inspection, and reduced maintenance and handling activities. As compared to single unit, the integrated approach requires an advanced stage of synthesis. Thus, in general, it will be more intricate than single unit processes. Nonetheless, it could give terms of lower cost and simplified manufacturing. This chapter is intended to recapitulate current works on renewable and sustainable materials produced from either single unit process or integrated approach through chemical or biological methods or combination of these methods for various green technology applications.

M. F. Yhaya et al., *Renewable and Sustainable Materials in Green Technology*, SpringerBriefs in Applied Sciences and Technology,
https://doi.org/10.1007/978-3-319-75121-4_4

4.2 Energy Applications

Fossil and nuclear energy are both non-renewable and non-sustainable. Although hydroelectricity energy is renewable, it is not sustainable due to the flooding of forests. Energy demand is a major challenge of the twenty-first century. It is projected that the world's energy demand will keep increasing and reach up to 50% and more by 2030 in all sectors as a result of escalating growth in population and rapid technological advancement (Ahmad and Riffat 2015). In meeting this growing trend, new and reliable energy sources in a sustainable and renewable manner are compulsory to the economic and social fabric of nations.

A plethora of new and improved materials is pivotal in meeting this demand of transition stage to a sustainable energy system. This includes the materials for harvesting energy from renewable sources, transferring energy, storing energy and/or converting it into other energy forms. Therefore, in an age when the demand for traditionally exploited natural resources is overtaking supply, the search for renewable and sustainable materials towards energy applications should be prioritised. This section discusses a few energy applications based on renewable and sustainable materials.

4.2.1 Biofuels as Alternative Energy Supply

The accelerating trend in global energy demand as a result of emerging economies and the recent increases of oil prices have led to growing numbers of research and innovation in biofuels. Biofuels are perceived as an alternative that able to replace fossil fuels in short term (Chisti 2007). Biofuels consist of energy derived from living organisms. They can be generated from starch, vegetable oils, animal fats, waste biomass, algal biomasses or microbes (Song and Shi 2008). These sources are non-toxic, biodegradable and can be replenished. In general, there are three main types of biofuels, which are ethanol, biodiesel and biojet fuel. Among the first biofuels used by human were wood and charcoal. Later, bioethanol is obtained from anaerobic fermentation of sugars by yeast while biodiesel is obtained from alcoholysis of vegetable or animal oils/fats.

According to UNEP (2009), biofuels can be categorised into first-, second-, third- and fourth-generation biofuels based on feedstock used and availability. First-generation biofuels originated from sugars and vegetable oils from crops. Second-generation biofuels are extracted from biomasses, mostly from agricultural waste. The third one is based on algae. Algae contain carbohydrates and lipids (fats) that can be converted into bioethanol and biodiesel, respectively. Fourth-generation biofuels are produced by synthetic photosynthetic process using microorganisms. Only the first generation is discussed in detail due to them are still the largest in terms of biofuels quantity produced.

Looking at the chemical approach point of view as previously discussed in Chap. 2, in the industry, ethanol can be synthesised by reacting ethane (ethylene)

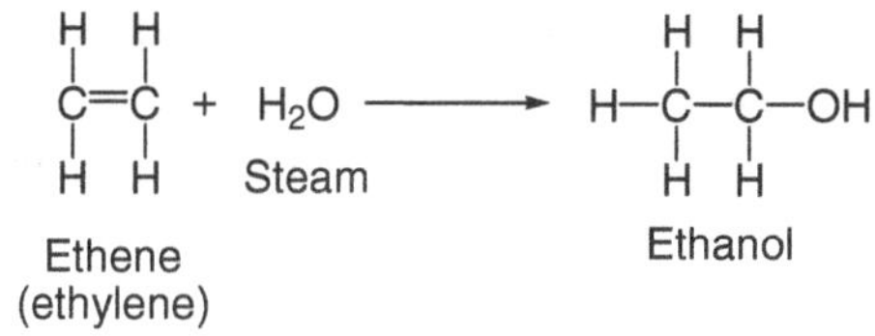

Ethene (ethylene) + H_2O (Steam) ⟶ Ethanol

Fig. 4.1 Ethanol production using ethane derived from petroleum

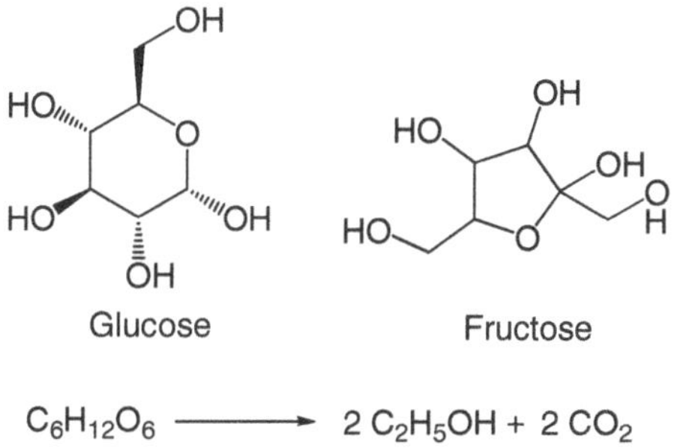

Glucose   Fructose

$$C_6H_{12}O_6 \longrightarrow 2\,C_2H_5OH + 2\,CO_2$$

Fig. 4.2 Fermentation of glucose and fructose into ethanol and carbon dioxide

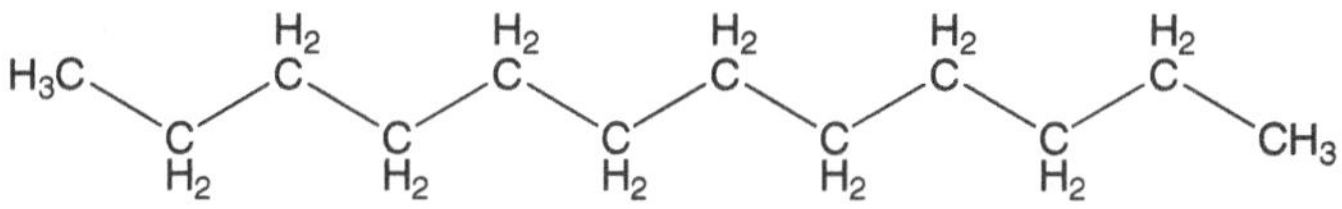

Fig. 4.3 Average chemical formula for petroleum-based diesel is C_{12}

with steam (Fig. 4.1). Meanwhile, bioethanol is produced by fermentation of sugars. Splitting off sucrose molecule will form glucose and fructose, which will undergo fermentation to produce ethanol and carbon dioxide (Fig. 4.2). The pure ethanol will then be obtained by distillation.

Petroleum-based diesel consisting mainly of C_{10}–C_{15} with the average of C_{12} (Fig. 4.3). The process involves cracking down the petroleum in the distillation tower.

Meanwhile, biodiesel can also be obtained by transesterification of vegetable oil. Vegetable oils are mostly triglycerides, which their basic structure is a combination of three molecules of fatty acids and a molecule of glycerol, such as palm oil (Fig. 2.4). Palm oil consists mainly of palmitic, oleic and linoleic acids. From Fig. 2.4, it can be seen that fatty acids have long alkyl chains, similar to petroleum-based diesel. Due to the long carbon chains, incomplete combustion of diesel will produce more soot as compared to petrol (gasoline).

The usage of food crops to produce biofuels, however, has created food versus fuel controversy. In order to solve this problem, source from non-edible crops such as Jatropha may be used. Jatropha is well known for its ability to survive in barren environment. Its seeds can provide triglycerides for conversion into biodiesel.

However, it must be remembered that as with other crops, Jatropha requires plenty of water and nutrient to thrive.

Despite the controversy of using food crops to produce biofuels, these are advantages and disadvantages of using them.

Advantages

- Renewable since the crop can be planted many times.
- Reducing total dependency on non-renewable resource such as petroleum.
- Reducing greenhouse gases and pollution as compared to fossil fuels.
- Economic security since not all countries are blessed with petroleum.

Disadvantages

- The crop planting requires plenty of water and fertilisers. In some places, the water source is very limited while fertilisers may pollute underground water and water bodies.
- Biofuels may reduce our dependence on fossil fuels but will still polluting the air, as they will eventually end up as smoke.
- In agriculture, usually, the crop rotation is practiced to increase soil fertility. Continuous monoculture will soon deprive the soil the nutrients it needed.
- Using food as source of biofuel will rise up its price. This is inappropriate since some people are still starving elsewhere in this world. Cassava is an interesting candidate for biofuel purpose since it is starch crop with the highest energy content per acre. It requires almost no fertiliser or irrigation, easy to culture and grow.

Throughout literature, most recent and thorough review of biofuels from microalgae as promising sustainable and renewable materials for energy supply is presented in Shuba and Kifle (2018) by emphasising the benefit and various forms of these biofuels, their recent development and production through genetic and metabolic engineering as well as prospect and promises and challenges in this industry. On the other hand, biofuels research also aims to produce alcohol in terms of ethanol. There are two major approaches utilised in this production which are direct fermentation and indirect fermentation.

Biofuels also can be produced via biological approach based on microbes. With this regard, Elshahed (2010) provided an overview of microbiological aspects of biofuel production together with state of the art, economic viability and future direction of this area as renewable energy sources. On the other hand, Tajarudin et al. (2016) also discussed energy production by microorganism through biological process such as bacteria and algae. In general, from the reports in the open literature, we can say that production of biofuels offers benefits in terms of reduction of greenhouse gas and its ability to replace fossil fuels. However, concerns regarding the overall sustainability of biofuels have been raised in relation to food production, water consumption and other resources to generate biomass. Taking this into account, thus a thorough impact analysis on sustainability of these materials by considering environmental, social and economic perspectives should be taken in the future.

Fig. 4.4 Heat exchangers as the auxiliary system to the air conditioner. **a** = hot and humid external air, **b** = cool and dry air to the air conditioner, **c** = hot and humid internal air from air conditioner, **d** = hotter and even more humid air towards outside and **e** = the heat exchanger module made of polymeric membrane-based heat and mass transfer surfaces

4.2.2 Polymeric Membrane-Based Heat Exchangers for Energy Recovery Applications

Heat exchangers are the heart of energy recovery system which often used to recover waste energy either for industrial or building applications. The energy recovery system is also called as energy recovery ventilator which is known as one of energy-efficient technologies and has been used in various climatic conditions (Rasouli et al. 2010; Bao et al. 2016). Thorough reviews on the work, mechanism, principles and application of this system are reported in several works in the open literature (Cuce and Riffat 2015; Ahmad and Riffat 2012, 2013).

In previous time, heat exchangers were originally made of metal to transfer sensible heat but with the demand of using more sustainable and renewable materials and the needs of transferring latent heat, their surface area has been replaced with polymers and their composites. The composites are made of treated natural fibres forming membrane sheets and stacked up as a module (Fig. 4.4). A thorough review of polymeric membrane-based heat exchangers is presented by Yhaya (2016) with a major discussion on effect of chemistry and chemical composition to their performance. The major advantage of the polymeric membrane-based heat exchangers is that they can be used to recover energy in hot-humid area in which they are placed to be integrated with the air conditioning system for energy recovery applications energy based on heat and mass transfer mechanism (Ahmad et al. 2016). The development of these heat exchangers consists of porous fibres impregnated with polymer-based solution that able to capture moisture from the hot and humid airstream.

In this system, heat is transferred and moisture is absorbed by the membrane surfaces of the heat exchanger, and thus, temperature is reduced as much as possible which often depends on its ability and performance factors before entering air conditioning system. The humid and hotter air from inside is blown from inside, carrying the heat and moisture, and drying up the module as it is blown out to the environment (Yang et al. 2015). As a result, the air conditioning system workload is reduced and so does the energy consumption.

4.3 Building and Infrastructure Applications

Green buildings and green infrastructures are two important criteria for solving urban and climatic challenges by built environment or construction sector with nature. In response to these criteria, careful selection of environmentally renewable and sustainable materials is essential for designers, architects or engineers to start incorporating sustainable concept in their building and infrastructure designs. However, the extensive implementation of the criteria is hindered by several significant challenges and barriers (Balaban and de Oliveira 2017; Akadiri 2015). Recently, bio-based building materials are becoming popular and used in numerous fields and various applications such as polymers, fillers, coatings, adhesives, impregnation materials or high-performance composites. For instance, Mija et al. (2017) explored a study on humins, a class of organic compound or biomacromolecular by-product of a hydroxymethylfurfural/furandicarboxylic acid biorefinery as promising material for producing sustainable carbohydrate-derived building materials. These materials potentially can be used as a renewable and sustainable source for the production of composite and wood impregnation.

4.3.1 Renewable Self-Healing Materials for Structural Application

From the material point of view, repair and maintenance of building materials or infrastructure are needed to increase service life and prevent cracks from propagating. This can be made by using numerous high-quality materials, but how many are considered as renewable and sustainable? Therefore, an emerging concept of self-healing agents or materials can be a possible solution for this. Self-healing materials are defined as substances that pose the built-in ability to self-repair internal damage all by themselves without any external diagnosis.

The design of materials with healing ability has attracted attention for various structural applications within buildings and infrastructure (Dry 2000; Qian et al. 2009; Van Tittelboom et al. 2011). However, most of these approaches are incorporating an expansive and non-renewable component in the structural form such as concrete which starts to expand and fill voids and cracks when triggered by carbonation or moisture ingress (Hosoda et al. 2007; Sisomphon et al. 2011). Then, a simplistic method to produce a durable and self-healing superhydrophobic surface for building materials is studied by Zulfiqar et al. (2017) using acetone treatment.

Later, the demand of switching from non-renewable materials to renewable materials has attracted attention on the utilising of bacteria as a promising technique to stimulate the self-healing mechanism. A bio-based self-healing agent by bacteria as renewable and sustainable materials is presented in Gupta et al. (2017) by addressing the effectiveness characteristics of bacterial self-healing, research gaps and scope of future research work. It is evidently seen that self-healing materials for building and infrastructure applications offer considerable practical advantages, as they would allow overcoming the difficulties associated with internal damage diagnosis and repair.

4.3.2 Natural Fibres and Wastes for Building Thermal Insulation

Building insulation is used for thermal purpose which commonly done using materials obtained from mineral or petrochemical-based insulation such as polystyrene. These materials cause significant harmful effects on the environment due to their production stage, in terms of the use of non-renewable materials, fossil energy consumption and to the disposal stage (Asdrubali et al. 2015). With the introduction of the concept of sustainability in building design as well as green building requirements and codes, researchers and designers have begun to develop thermal insulating materials based on natural sources or recycled materials which can be replenished and renewed. These natural fibres can be plant-based and animal-based such as wood fibre, kenaf, wool, hemp, cotton and flax. These natural fibres are processed into loose, semi-rigid or rigid sheets. For building applications, these sheets are usually treated with fire retardants. The natural fibre insulation is also capable of providing acoustic insulation by absorbing sound and preventing reverberation. High specific heat storage capability of natural fibres can be exploited to reduce overheating. These insulation sheets may be located underneath the roof.

Both synthetic and natural insulation have their advantages as well as their disadvantages. Synthetic fibres made from silica and asbestos are very harmful to the lungs. As compared to their synthetic counterparts, natural fibres do not require the installer to wear protective clothes. So, if used appropriately, these materials can deliver thermal insulation comparable to other insulation materials, but with a lesser or potentially reduce carbon emission and lower health problems during installation. Manufacturing of synthetic insulation is releasing more carbon dioxide into the air instead. The offcuts after the installation of natural fibre insulation can be sent directly for composting while synthetics ones have to be sent to landfill. However, lignocellulosic natural fibre-based insulation is susceptible to water damage and has shorter lifespan.

A recent review of developments on the building thermal insulations is reported in Aditya et al. (2017). The reviews also cover life cycle analysis and potential emission reduction by using appropriate building insulation materials. Apart from natural fibres, Sagbansua and Balo (2017) presented a work on novel biocomposite materials obtained from different natural wastes (fly ash) and natural (modified corn oil, perlite and clay) raw materials which potentially suitable as sustainable insulation materials for building components. On the other hand, Liuzzi et al. (2017) discussed a state of the art of several building thermal insulation products made of agro-wastes such as cereal straw, hemp and olive waste. It is scientifically evident that the development and the application of bio-based building insulation materials significantly able to reduce the environmental impacts of buildings minimising the energy demand during the construction and operation phases (Dahy 2017). However, in the future, more works should be carried out on potential of natural sources as building insulation that could meet the requirements of sustainability in building sector gearing towards development of prototype and commercialisation.

4.3.3 Biocomposites for Building Envelope

Building envelope is the physical components between the interior and exterior parts of a building. These components typically cover walls, floors, roofs, fenestrations and doors or any opening structures such as windows, skylights and clerestories. In the age of 'going green', good building envelope should use environmentally and sustainable exterior wall materials and designs that are aesthetically pleasing, well ventilated and climate appropriate.

From the open literature, it is found that a significant interest in the use of plant fibres as biocomposite for building materials is growing rapidly (Khoshnava et al. 2017; Gurunathan et al. 2015; Khalil et al. 2012). Biocomposite is a term used to indicate fibre-reinforced polymer composite materials in which the matrix or fibres are bio-based. These include jute, flax and hemp which offer reduction of weight, added functionality and occupational health advantageous. These materials have gained interest in recent years especially in building sector prompted by the increased of awareness and drive towards green technologies.

Liu et al. (2017) recently have presented a review of hemp, a natural plant cellulose fibre as biocomposite material due to its potential properties of high strength. From the review, it is clear that the suitability of this material is due to the advantageous properties of its characteristics in terms of biodegradability, economically, high stiffness and strength, and low density. Usage of biocomposite for interior building façade is reported by Roig (2017). The material is not just durable and sustainable which allows higher design freedom to designers, but able to prove thermal and acoustic insulation as well as improve indoor air quality in new buildings and restorations providing a breathable construction system.

4.4 Environmental Applications

Materials and their usage can have an enormous environmental impact in various aspects from sourcing, extraction, processing to end of life or in other words from 'cradle to grave'. So, on the environmental front, the usage of renewable and sustainable materials plays a pivotal part to meet the increasing demands from society on pollution remediation and green products. This section presents a review of selected renewable and sustainable materials used for several environmental applications.

4.4.1 Sustainable Activated Carbon for Water and Wastewater Treatment

Water is the liquid of life. Despite water covering the vast majority of Earth's surface, only less than 1% is usable. The rest is either saline (sea) water or trapped in icecaps and glaciers. As of December 2017, United Nations (UN) estimates that the current world population is 7.6 billion. The ever-increasing human population is putting the stress into fresh water supply. In the Middle East countries, desalination is used to turn seawater for public use. Since 1998, Singapore has been turning sewage into potable water called NEWater. In Malaysia, the rivers were dammed causing flooding of forests and displacement of wildlife. In India, groundwater extraction keeps on increasing and is not sustainable anymore. Some of the residents are using water from the river polluted with human faeces and dead bodies. African countries are also facing freshwater shortage. Even the developed countries are not spared. In order to treat huge volume of water in a very short time, processing of raw water into potable water may be done using activated carbon.

Activated carbon is a processed carbon with very small macro/micropores with high surface area for adsorption (Gaspard et al. 2014). It can be produced from carbon compounds; the cheapest source would be lignocellulosic waste, which later is turned into charcoal. The charcoal can be activated either physically or chemically, but the latter is better as the temperature used and time taken to activate are much lower. The activation actually means unplugging the pores, leaving the cavities ready for adsorption. The activated carbon is used for separation, filtration, purification and medical uses. Since most activated carbon is in powder form, its recovery can be a hassle. Activated carbon fabric (ACF) is much more convenient since it is in cloth/sheet form, made from viscose rayon, polyvinyl chloride (PVC), phenolic resin, polyacrylonitrile (PAN) and pitch precursors (Marcuzzo et al. 2013). Reactivation of activated carbon by thermal is the most common but expensive. Reactivation means desorption, which is the reverse of adsorption.

4.4.2 Polymers and Biopolymers for Water Retaining and Soil Stabilisation

Polymers and biopolymers have been used comprehensively for environmental applications. They have been known as one of green technological options. A comprehensive review of the environmental applications of biopolymers is presented in Kanmani et al. (2017), by providing a fresh insight into the future prospects of research in this area. These materials have been identified as viable soil conditioners due to their ability to stabilise surface structure of soil as continuity of pore (Ayelden et al. 2017).

In the previous decade, water-soluble polymers were recognised to prevent erosion in soil. Polyacrylamide is a long-chain synthetic polymer that acts as a strengthening agent, binding soil particle together, and holding soils in place. It is found to be effective in enhancing the stability of soil aggregates and increase soil infiltration of sandy loam soils. In this regard, Maghchiche et al. (2010) conducted a research to determine the effects of different synthetic polymers and biopolymers at low concentration (0.03–1%) for water retaining and soil stabilisation in arid and semi-arid regions. It was found in their study that the polymer composites (10 mg/L polyacrylamide and 0.5 g/L cellulose) in soil could improve better soil physical characteristics and augment 60% of water retention at arid soils compared with application of any other polymer at the same concentration. They also stated that by using low concentration of biopolymers from plant fibres and polymers from synthetic plastics compounds or wastes plastic industry could help to improve the productivity of the soils.

On the other hand, Hataf et al. (2018) investigated the potential of clay soil stabilisation by using chitosan biopolymer at different concentrations synthesised from shrimp shell waste. From the study, it was found that the incorporation of the biopolymer helped to increase the inter-particle interaction soil particles which led to improved mechanical properties. In earlier study, a series of biopolymers added to irrigation water were tested to reduce erosion-induce soil losses in Orts et al. (2000) based on laboratory conditions. Throughout these studies, it is appeared that the usage of biopolymers offers a renewable, environmentally friendly, inexpensive material. Nevertheless, more studies should be performed in this domain to look for potential of different polymers or biopolymers for water retaining and soil stabilisation in different climatic regions.

4.4.3 Lactic Acid as Versatile and Environmental Friendly Chemical Resources

Production of versatile chemical resources from renewable and sustainable materials have gained considerable highlight in recent years. In this regard, production of lactic acid from bacteria is seen as an option that also able to cope with environmental problems and cost-effectiveness. Lactic acid is an important organic acid used in the chemical, pharmaceutical, cosmetics, agriculture and food industries. Apart from its versatile application in these industries, lactic acid also can be a potential raw material for the production of biodegradable plastics (Karamanlioglu et al. 2017).

The production of lactic acid by biotechnological is preferable to chemical process due to the possibility of obtaining pure isomers which can be easily assimilated in human body. Besides, this material is also important in polymerisation process to produce fibres and films. Wang et al. (2015) reported the development in biotechnological production of lactic acid produced and its potential for various processes. In another review, Andreev et al. (2018) stated that lactic acid fermentation helped to reduce the amount of pathogens and to minimise the nutrient loss and thus increasing agricultural value of plants.

Fig. 4.5 Self-watering system. Due to the negative pressure and careful design of the membrane, the water is absorbed through the column filled with treated natural fibre/synthetic membrane. As the water travels up the column, the suspended solids will be filtered out due to gravity. Water will eventually reach the top, dripping out and watering the plants

Thus, with the development of new bioreactors configurations, use of renewable resources and downstream processing, the production of lactic acid can be improved for various green technology applications.

4.4.4 Water-Permeable Polymer for Self-Watering System

Mimicking the nature can be useful for solving problems. A tree is one of the great natural creations because it can transport water from the soil into the top of the tree, fighting gravity using transpiration of water from leaves. The water is capable to be transported up as high as 115 m as in coast redwood (*Sequoia sempervirens*), the world's tallest tree (Mason Earles et al. 2016).

Using similar concept, a ***water-permeable polymer for self-watering system*** can be developed (Fig. 4.5). This system may be used in some rural areas where rivers are far away and electricity supply is non-existent. The pathway for the water travelling should not be straight upwards as the gravity will pull it back down. The water will cover much longer distance before reaching the top of the column. Some of the key design points learned from the trees are that the space inside the membrane must be very small, plus the space must be capable to hold the water temporarily as it travels upwards.

References

L. Aditya, T.M.I. Mahlia, B. Rismanchi, H.M. Ng, H.B. Aditiya, A review on insulation materials for energy conservation in buildings. Renew. Sustain. Energy Rev. **73**, 1352–1365 (2017)
M.I. Ahmad, S. Riffat, Review on heat recovery technologies for building applications. Renew. Sustain. Energy Rev. **16**(2), 1241–1255 (2012)
M.I. Ahmad, S. Riffat, Review on physical and performance parameters of heat recovery system for building applications. Renew. Sustain. Energy Rev. **28**, 174–190 (2013)
M.I. Ahmad, S. Riffat, Building energy consumption and carbon dioxide emissions: threat to climate change. J. Earth Sci. Climatic Change. 1–3 (2015). http://dx.doi.org/10.4172/2157-7617.S3-001
M.I. Ahmad, F.Z. Mansur, S. Riffat, Applications of air-to-air energy recovery in various climatic conditions, in *Renewable Energy and Sustainable Technologies for Building and Environmental Applications* ed. by M.I. Ahmad, M. Ismail, S. Riffat (Springer International Publishing, 2016)
P.O. Akadiri, Understanding barriers affecting the selection of sustainable materials in building projects. J. Build. Eng. **4**, 86–93 (2015)
N. Andreev, M. Ronteltap, B. Boincean, P.N.L. Lens, Lactic acid fermentation of human excreta for agricultural application. J. Environ. Manage. **206**, 890–900 (2018)
F. Asdrubali, F. D'Alessandro, S. Schiavoni, A review of unconventional sustainable building insulation materials. Sustain. Mater. Technol. **4**, 1–17 (2015)
M. Ayeldeen, A. Negm, M. El-Sawwaf, M. Kitazume, Enhancing mechanical behaviours of collapsible soil using two biopolymers. J. Rock Mech. Geotech. Eng. **9**(2), 329–339 (2017)
O. Balaban, J.A.P. de Oliveira, Sustainable buildings for healthier cities: assessing the co-benefits of green buildings in Japan. J. Cleaner Prod. **163**, 68–78 (2017)
L. Bao, J. Wang, H. Yang, Investigation on the performance of a heat recovery ventilator in different climate regions in China. Energy **104**, 85–98 (2016)
Y. Chisti, Biodiesel from microalgae. Biotechnol. Adv. **25**(3), 294–306 (2007)
P.M. Cuce, S. Riffat, A comprehensive review of heat recovery systems for building applications. Renew. Sustain. Energy Rev. **47**, 665–682 (2015)
H. Dahy, Biocomposite materials based on annual natural fibres and biopolymers—design, fabrication and customized applications in architecture. Constr. Build. Mater. **147**, 212–220 (2017)
C.M. Dry, Three designs for the internal release of sealants, adhesives, and waterproofing chemicals into concrete to reduce permeability. Cement and Concrete Research, **30**(12), 1969–1977 (2000)
M.S. Elshahed, Microbiological aspects of biofuel production: current status and future directions. J. Adv. Res. **1**(2), 103–111 (2010)
S. Gaspard, P.C. Nady, D. Axelle, C. Thierry, J.R. Valerie, Activated carbon from biomass for water treatment. RSC Green Chem. Ser. **25**, 46–105 (2014)
S. Gupta, S.Z. Pang, H.W. Kua, Autonomous healing in concrete by bio-based healing agents—a review. Constr. Build. Mater. **146**, 419–428 (2017)
T. Gurunathan, S. Mohanty, S.K. Nayak, A review of the recent developments in biocomposites based on natural fibres and their application perspectives. Compos. A Appl. Sci. Manuf. **77**, 1–25 (2015)
N. Hataf, P. Ghadir, N. Ranjbar, Investigation of soil stabilistion using chitosan biopolymer. J. Cleaner Prod. **170**, 1493–1500 (2018)
T. S. Hosoda, T. Kishi, Self-repairing function for cracks of SCC with expansive agent. in *Proceedings of the Second International Symposium on Self-Compacting Concrete*, 483–490 (2001)
P. Kanmani, J. Aravind, M. Kamaraj, P. Sureshbabu, S. Karthikeyan, Environmental applications of chitosan and cellulosic biopolymers: a comprehensive outlook. Biores. Technol. **242**, 295–303 (2017)
M. Karamanlioglu, R. Preziosi, G.D. Robson, Abiotic and biotic environmental degradation of the bioplastic polymer poly(lactic acid): a review. Polym. Degrad. Stab. **137**, 122–130 (2017)
H.P.S.A. Khalil, I.U.H. Bhat, M. Jawaid, A. Zaidon, Y.S. Hadi, Bamboo fibre reinforced biocomposites: a review. Mater. Des. **42**, 353–368 (2012)

S.M. Khoshnava, R. Rostami, M. Ismail, A.R. Rahmat, B.E. Ogunbode, Woven hybrid Biocomposite: Mechanical properties of woven kenaf bast fibre/oil palm empty fruit bunches hybrid reinforced poly hydroxybutyrate biocomposite as non-structural building materials. Constr. Build. Mater. **154**, 155–166 (2017)

M. Liu, A. Thygesen, J. Summerscales, A.S. Meyer, Targeted pre-treatment of hemp bast fibres for optimal performance in biocomposite materials: a review. Ind. Crops Prod. **108**, 660–683 (2017)

S. Liuzzi, S. Sanarica, P. Stefanizzi, Use of agro-wastes in building materials in the Mediterranean area: a review. Energy Procedia **126**, 242–249 (2017)

Maghchiche, H.B. Immirzi, Use of polymers and biopolymers for water retaining and soil stabilization in arid and semiarid regions. J. Taibah Univ. Sci. **4**, 9–16 (2010)

J.S. Marcuzzo, O. Choyu, A.P. Heitor, O. Satika, Influence of thermal treatment on porosity formation on carbon fiber from textile PAN. Mater. Res. (Sao Carlos, Braz.). **16**, 137–44 (2013)

J. Mason Earles, O. Sperling, L.C.R. Silva, A.J. McElrone, C.R. Brodersen, M.P. North, M.A. Zwieniecki, Bark water uptake promotes localized hydraulic recovery in coastal redwood crown. Plant, Cell Environ. **3**(9), 320–328 (2016)

A. Mija, J.C. van der Waal, J.M. Pin, N. Guigo, E. de Jong, Humins as promising material for producing sustainable carbohydrate-derived building materials. Constr. Build. Mater. **139**, 594–601 (2017)

National Research Council (NRC), Unit manufacturing processes: issues and opportunities in research (National Academies Press, Washington, D.C., United States, 1995)

W.J. Orts, R.E. Sojka, G.M. Glenn, Biopolymer additives to reduce erosion-induced soil losses during irrigation. Ind. Crops Prod. **11**(1), 19–29 (2000)

S. Qian, J. Zhou, M.R. De Rooij, E. Schlangen, G. Ye, K. Van Breugel, Self-healing behavior of strain hardening cementitious composites incorporating local waste materials. Cement and Concrete Composites. **31**, 613–621 (2009)

M. Rasouli, C.J. Simonson, R.W. Besant, Applicability and optimum control strategy of energy recovery ventilators in different climatic conditions. Energy Build. **42**(9), 1376–1385 (2010)

I. Roig, Biocomposites for interior facades and partitions to improve air quality in new buildings and restorations. Reinf. Plast. (2017). https://doi.org/10.1016/j.repl.2017.07.003

L. Sagbansua, F. Balo, A novel simulation model for development of renewable materials with waste-natural substance in-sustainable buildings. J. Cleaner Prod. **158**, 245–260 (2017)

E.S. Shuba, D. Kifle, Microalgae to biofuels: 'promising' alternative and renewable energy, review. Renew. Sustain. Energy Rev. **81**(1), 743–755 (2018)

K. Sisomphon, O. Copuroglu, A. Fraaij, Application of encapsulated lightweight aggregate impregnated with sodium monofluorophosphate as a self-healing agent in blast furnace slag mortar. Heron **56**(1/2), 13–32 (2011)

D.J. Song, F.D. Shi, Exploitation of oil-bearing microalgae for biodiesel. Chin. J. Biotechnol. **24**(3), 341348 (2008)

H.A. Tajarudin, M.R. Tamat, M.F. Othman, N.A. Serri, N.Q. Zaman, Energy recovery by biological process, in *Renewable Energy and Sustainable Technologies for Building and Environmental Applications* ed. by M.I. Ahmad, M. Ismail, S. Riffat. (Springer International Publishing, 2016)

UNEP. Towards sustainable production and use of resources: Assessing biofuels (2009). http://www.compete-bioafrica.net/publications/publ/Assessing%20Biofuels-Summary-Web.pdf. Accessed on 1 Nov 2017

K. Van Tittelboom, K. Adesanya, P. Dubruel, P. Van Puyvelde, N. De Belie, Methyl methacrylate as a healing agent for self-healing cementitious materials. Smart Materials and Structures, **20**(12) (2011). https://doi.org/10.1088/0964-1726/20/12/125016

Y. Wang, Y. Tashiro, K. Sonomoto, Fermentative production of lactic acid from renewable materials: recent achievements, prospects, and limits. J. Biosci. Bioeng. **119**(1), 10–18 (2015)

P. Yang, L. Li, J. Wang, G. Huang, J. Peng, Testing for energy recovery ventilators and energy saving analysis with air-conditioning systems. Procedia Eng. **121**, 438–445 (2015)

M.F. Yhaya, Polymeric heat exchangers: effect of chemistry and chemical composition to their performance, in *Renewable Energy and Sustainable Technologies for Building and Environmental Applications* ed. by M.I. Ahmad, M. Ismail, S. Riffat (Springer International Publishing, 2016)

U. Zulfiqar, M. Awais, S.Z. Hussain, I. Hussain, T. Subhani, Durable and self-healing superhydrophobic surfaces for building materials. Mater. Lett. **192**(1), 56–59 (2017)

Chapter 5
Challenges, Future Outlook, and Opportunities

5.1 Current Limitations

Increasing threats to sustainability are leading both governments and industry to explore different criteria of the green economy including renewable and sustainable materials as well as green technology. As the green economy gains power, it is essential for all stakeholders to comprehend the important factors driving it, as they are relevant for policymaking and determining the future business models of industry. In addition, the demand and attention from many industries contributing to the development of industrial innovation have become increasingly considerable, consequential from the interrelationship between competitiveness, economic growth and the innovation itself.

The following section discusses the most significant barriers or limitations to the development of renewable and sustainable materials towards green technological option. These include cost and economic factors as well as technical and management in relation to the commitment towards innovation. In general, cost and economic factors would contribute to financial aspect of industry. Meanwhile, technology and management factors would often affect time needed for return on investment, risks and capability of industry to apply high technology.

5.1.1 Cost and Economic Factors

Cost and economic factors are among the critical criteria and main limitations that affect an investment's value particularly for implementation of new technology in industry. Various cost and economic factors need to be considered when determining and the current and expected future value or investment in any technological solutions. These include labour costs, interest rates, government policy and taxes.

M. F. Yhaya et al., *Renewable and Sustainable Materials in Green Technology*, SpringerBriefs in Applied Sciences and Technology,
https://doi.org/10.1007/978-3-319-75121-4_5

Therefore, in dealing with this circumstance, a proper procedure and analysis should be followed in ensuring the best approach is taken to achieve benefits while preserving savings. One of the main procedures to be followed is cost-benefit analysis, which is a systematic approach in predicting the strengths and weaknesses of options of new technology.

5.1.2 Technical and Management

The era of sustainability is motivated by both demand and supply sides. Looking at supply side, technical and management are important factors to be considered for the implementation of new technology in meeting the demand side. In this context, better technical support, sophisticated equipment and skilled workers are required. Thus, the momentum to make resource management as key priority is vital and should be in front of other components. However, the main constraint relates to this is hiring and training human resources to be skilled workers are hardly difficult and quite time-consuming. Therefore, planning and readiness are two important elements which should be taken into account before execution of any projects or implementation of technological solutions particularly involving green technology and sustainability.

The skilled workers performing manual jobs can be produced through experience and obviously, their wages would be higher than normal workers. However, normal workers could always be temporarily replaced to keep the production ongoing. However, the frequently replacement position of skilled worker to normal worker will increase a critical piece of manufacturing technology to fail then consequently the industry will start to lose money right away, in terms of production. This would cause the industry faces various problems in relation to delivery deadlines and eventually gives a bad reputation and image. On the other hand, implementation of new technology also can introduce new and dangerous workplace where the new standard operating procedures are required in term hazards and physical safety. Some of innovations produce wastes during the process of production. Therefore, the consideration of industrial waste management needs to be revised.

5.2 Fossil Fuels and Their Alternatives

In these tough times, it is imperative to find alternatives for current resources. As everyone knows, the rates of regeneration of fossil fuels are very slow as compared to their consumption. Since energy is important to get things around and to move the world, their replacement must be established soon.

5.2.1 Fossil Fuels and Global Warming

As mentioned previously in Chap. 2, petroleum, natural gas, coal and peat are classified as fossil fuels. It takes years and millions of years to regenerate fossil fuels. Apart from being non-renewable resources, the burning of fossil fuels is also generating carbon dioxide (a greenhouse gas), contributing to the global warming. Global warming (climate change) is associated with the melting of icecaps and icebergs, the drowning of cities due to the increase of sea level, unpredictable seasons, droughts, floods, ocean acidification and species extinctions.

Petroleum It is the most popular fossil fuel, formed by fossilisation of zooplankton and algae under intense heat and pressure. Russia is the world's largest producer of petroleum, followed by Saudi Arabia and the United States. The world's largest consumer of petroleum are Asian countries (China, Japan and India), followed by the United States and Russia. Petroleum is mined either on land or offshore. The occurrence of natural petroleum springs is very rare. After extraction, crude petroleum is refined using distillation tower, where the hydrocarbons in the petroleum are cracked (broken down) and separated. The bottom of the tower will produce very viscous black fluid called bitumen for road pavement and insulation. The top of the tower will produce colourless cooking gas, while the middle part will produce petrol, kerosene, diesel and raw chemicals for making pharmaceuticals and synthetic polymers.

Natural gas The world is believed to still possess large resources of natural gas, unlike petroleum. Usually, it is found together with other fossil fuels such as petroleum and coal. It consists mainly of methane (CH_4), formed when layers of decomposing plant and animal matter are exposed to intense heat and pressure underground for over millions of years. It is burned for cooking, rural heating during winter, and electricity generation. In the petroleum reserve, the natural gas is situated on top of the former, only to be burnt as flare gas that is the trademark of oil rigs. Due to its low density, natural gas is usually turned into liquefied natural gas (LNG) using very high pressure before transportation. The release of natural gas (mostly methane) into the environment is contributing to the global warming, as methane is 83 times more potent as compared to carbon dioxide. Combustion of natural gas will produce carbon dioxide, while incomplete combustion will release carbon monoxide, a lethal gas. Some death cases were reported during the winter due to the leakage of odourless carbon monoxide.

Coal It is a combustible black or brownish-black rock made of terrestrial plants subjected to intense pressure and heat. Coal mining can be done on both the ground surface and underground. Underground mining is more risky due to flooding, cave-ins, explosion, chemical leakage and electrocution. China is both the main producer and consumer of coal, followed by the United States and India. Coal consists mainly of carbon and also contains hydrogen, oxygen, nitrogen and sulphur. The coal with the highest carbon content (anthracite) is the best and the cleanest. During early times, coal was burned to heat the water and turned it into steam, which later turned into mechanical energy to move the trains and ships. Coal is still currently in use to generate electricity in many countries. Burning of coal will release carbon dioxide, sulphur dioxide (SO_2), nitrogen oxides (NOx) and particulate matters (PM). Carbon dioxide contributes to the global warming, while sulphur dioxide and nitrogen oxides will cause acid rain, ground-level ozone and smog. Particulate matters are harmful to human, affecting respiratory, cardiovascular, and nervous systems. However, coal (as coke) is used as reducing agent for production of steel and other iron-based products. Some valuable chemicals such as ethanoic acid, acrylic acid, olefins, formaldehyde, ammonia and urea can be produced from coal. Ammonia and urea can be turned into fertiliser.

Peat In 1992, United Nations Framework Convention on Climate Change (UNFCCC) classified peat as non-renewable resource. It is formed from partly decayed vegetation or organic matter under acidic and anaerobic conditions. However, the regeneration of peat is very slow, about 1 mm/year and damaged bogs may take 100 years to regrow. Peat itself is infertile but it is able to store many nutrients required for plants to grow. Peat land is an effective carbon sink. The reclamation of peat lands for other purposes will release the carbon dioxide back into the atmosphere. Most of the harvested peat is burned as fuel, with few percentage of it is used for sewage filtration. Similar to coal, the burning of peat will produce small particulate matters. Unfortunately, an uncontrolled peat fire may smoulder undetected underground and capable of spreading very fast. The most remembered is the 1997 Southeast Asian haze that originally started as lignocellulosic biomass burning by Indonesian farmers. After reaching the flammable peat-swamp areas, the fires are too difficult to control, spreading the particulate matters to the neighbouring countries, causing very poor atmospheric visibility and health problems. The agriculture, transportation and tourism industry were badly hit. The children and the elderly were those most affected by this haze. In 2010, summer temperature of 40 °C self-ignited large peat deposit in Moscow and many were left homeless due to the ravaging fire.

5.2.2 Natural Resources and Ecosystem

Natural resources can be defined as resources that have been existing in this world without human interventions. Meanwhile, an ecosystem can be defined as a community of living organisms in conjunction with their physical environment. To reduce human dependence on the hydroelectricity and fossil fuels, the scientists, technologists, and engineers around the world are looking into alternatives. Alternative energy must be sought with very little or no disturbances to the ecosystem. **Biofuels** are already described in detail in Chap. 4. Despite its huge contribution to some countries, nuclear energy is not discussed here. Although the uranium is renewable, the one used for nuclear fusion is the rare ones and the nuclear waste disposal is still a major environmental and health concern. Nuclear disasters such as the Hiroshima and Nagasaki bombing (1945) and Chernobyl (1986) still serve as a stark reminder of the dangers of nuclear energy.

Methane from Anaerobic Decomposition Anaerobic decomposition of organic materials by methanogens will produce methane. There should be a systematic way of collecting all the methane gas generated from sewage, municipal solid waste and lignocellulosic biomass decomposition. The methane gas then can be purified and/or liquefied. Currently, the methane gas is burned or released into the environment without much control.

Hydrogen as Fuel Cell Apart from fossil fuels and electricity, hydrogen (as in fuel cell) may also be used to propel the vehicles. Combustion or chemical reaction of hydrogen with oxygen will produce energy and water. However, the earlier way of producing hydrogen is not that sustainable. The water molecules were broken down into hydrogen gas and oxygen gas using electricity, which mainly originated from fossil fuel. To avoid the pitfall, the electricity used for electrolysis itself must be from renewable and sustainable resources. Chemically, bioethanol may be reacted with hot steam to produce hydrogen. Biologically, microbes such as green algae are cultivated for hydrogen production.

Solar Since the sun is not going anywhere anytime soon, harvesting of solar energy is the thing to do to gain quick energy. Solar panels may be located at strategic places for harvesting the sunlight. The heat collected will be converted into electricity for household and buildings use. At the moment, despite the ubiquitous solar car racing everywhere, majority of the cars are still running on fossil fuels. Even though the current technology limitation does not allow us to fully run a full-sized family car on solar energy, at least some of the sunlight should not go wasted. The roof should be the perfect place to instal the solar panel. The electricity generated can be used to run the electrical and air-conditioning system. This will reduce the dependence on fossil fuels as car air conditioner is operated using the car's engine.

Wind Wind mills should be installed in strategic locations to harvest the kinetic energy. This has long been practiced by farmers in the Netherlands to ground the grains and pumping the water out from lowlands so that the land could be farmed. However, the traditional wind mill design is always producing noise and shadow flicker from the rotating blades that can be regarded as environmental pollution. Currently, the bladeless wind turbine design is proposed which is less expensive to manufacture, totally quiet, and much safer for birds.

Wave and Tidal Energy For wave energy, the energy is captured from the sea waves. In contrast, tidal energy is driven by the gravitational pull of the moon and sun. Despite the differences, the key point in capturing these energies is that the mechanical work of seawater is captured and turned into electricity by devices.

Geothermal It is the thermal energy generated and stored in earth. The heat may be originated from deep down into the earth's core. Heat is conducted from the core to the surrounding rocks. Some of the rocks will melt to become magma. The magma, in turn, will heat the upper rock layer. If this is happening under the water catchment, it will eventually boil up and turn into hot water springs. If the boiling water is pushed out through columns in the rock, it will gushing out as geysers. For generations, these places are associated with leisure (bathing, self-healing and tourisms). Unites States generates more electricity from geothermal than any other countries in this world, but it is only a small percentage of total USA's energy supply.

Plants Although it is still at the early stage, scientists believe that electric can be generated directly from living plants or trees due to the different in pH between soil and the plants. The effect is similar to electric generated by putting two different metal electrodes on a lemon. However, in the case of plants, the electric is still generated even though the same electrodes are used. The electricity generation can be done without affecting the plant's growth in any way. At the moment, the current generated is just enough to power a small device to detect forest fires, monitoring environmental change and also the health of the plants themselves. In the future, it is expected that the electric harvesting may be upscaled to provide electricity for the community living in the rural areas or in the forest.

Bibliography

H. Bahl, W. Andersch, K. Braun, G. Gottschalk, Effect on pH and butyrate concentration on the production of acetone and butanol by *Clostridium acetobutylicum* growing in a continuous culture. Eur. J. Appl. Microbiol. Biotechnol. **14**, 17–20 (1982)

J. Ballongue, J. Amine, E. Masion, H. Petitdemange, R. Gay, Induction of acetoacetate decarboxylase in *Clostridium acetobutylicum*. FEMS Microbiol. Lett. **29**, 273–7 (1985)

O. Fond, G. Matta-Ammouri, H. Petitdemange, J.M. Engasser, The role of acids on the production of acetone and butanol by *Clostridium acetobutylicum*. Appl. Microbiol. Biotechnol. **22**, 195–200 (1985)

M.J. Johnson, W.H. Peterson, E.B. Fred, Oxidation and reduction relations between substrate and products in the acetone-butylalcohol fermentation. J. Biol. Chem. **91**, 569–91 (1931)

A.H. Kaksonen, B.M. Mudunuru, R. Hackl, The role of microorganisms in gold processing and recovery—a review. Hydrometallurgy **142**, 70–83 (2014)

O.P. Karthikeyan, A. Rajasekar, R. Balasubramanian, Bio-oxidation and biocyanidation of refractory mineral ores for gold extraction: a review. Crit. Rev. Environ. Sci. Technol. **45**(15), 1611–1643 (2015). https://doi.org/10.1080/10643389.2014.966423

D.H.P. Ng, A. Amit Kumar, B. Cao, Microorganisms meet solid minerals: interactions and biotechnological applications. Appl. Microbiol. Biotechnol. **100**, 6935–6946 (2016). https://doi.org/10.1007/s00253-016-7678-2

S. Philip, T. Keshavarz, I. Roy, Polyhydroxyalkanoates: biodegradable polymers with a range of applications. J. Chem. Technol. Biotechnol. **82**(3), 233–247 (2017)

M. Ponraj, A. Talaiekhozani, R. Mohamad Zin, M. Ismail, M.Z.A. Majid, A. Keyvanfar, H. Kamyab, Bioconcrete strength, durability, permeability, recycling and effects on human health: a review, in *Proceeding of the Third International Conference Advances in Civil, Structural and Mechanical Engineering-CSM* (2015)

D.E. Rawlings, D.B. Johnson, The microbiology of biomining: development and optimisation of mineral-oxidizing microbial consortia **153**, 315–324 (2007)

C. Rodrigues, L.P.S. Vandenberghe, A.L. Woiciechowski, J.C.R. de Oliveira, Soccol24-production and application of lactic acid. Curr. Dev. Biotechnol. Bioeng. 543–556 (2017)

J.S. Terracciano, E.R. Kashket, Intracellular conditions required for initiation of solvent production by *Clostridium acetobutylicum*. Appl. Environ. Microbiol. **52**, 86–91 (1986)

M. F. Yhaya et al., *Renewable and Sustainable Materials in Green Technology*, SpringerBriefs in Applied Sciences and Technology, https://doi.org/10.1007/978-3-319-75121-4

Index

M. F. Yhaya et al., *Renewable and Sustainable Materials in Green Technology*,
SpringerBriefs in Applied Sciences and Technology,
https://doi.org/10.1007/978-3-319-75121-4

Zeitfracht Medien GmbH
Ferdinand-Jühlke-Straße 7
99095 Erfurt, Deutschland
produktsicherheit@kolibri360.de